ENCYCLOPEDIA OF UNDERWATER INVESTIGATIONS

Second Edition

by Cpl. Robert G. (Bob) Teather C.V.
Royal Canadian Mounted Police

This book is dedicated to the memory of all public safety divers
who have given the ultimate sacrifice, to their families,
and to you who carry on the tradition.

Second Edition 2013
First Edition 1994

International Standard Book Number
ISBN 978-1-930536-72-2

Library of Congress Catalog Card Number
92-82800

BEST PUBLISHING COMPANY
2700 PGA Boulevard, Suite 104
Palm Beach Gardens, FL 33410
Printed in the USA

TABLE OF CONTENTS

FOREWORD

When this book was first published in 1994, it was a ground-breaking addition to diving literature. There was nothing like it in the field. To this day, it remains one of the pre-eminent publications in the area of underwater forensics and investigations. It was the brainchild of one man, Corporal Robert Gordon Teather, C.V., of the Royal Canadian Mounted Police (1947-2004).

Bob joined the RCMP in 1967 and was stationed in North Vancouver. While there, he served in Uniform Patrol and was acknowledged as a trained hostage taker-barricaded person negotiator and a forensic dive instructor. It was his drive and determination that vitalized the "E" Division dive team, which later was renamed the Underwater Recovery Team.

It was in the course of these duties that Bob assembled the knowledge base and developed the techniques discussed in this text. During the course of his work, Bob spoke regularly throughout Canada and the United States at a variety of conferences and to many public safety dive teams and other groups, including firefighters, policemen, physicians, dive agencies, and dive instructors. He would talk on topics that kindled his passions: underwater photography and documentation, underwater crime scene investigation, evidence recovery, body recovery, ice diving, and too many other topics to list. In fact, even prior to its original publication the content in this text was accepted as a court reference in superior court (USA) and coroner's court (Canada).

Bob gave selflessly during his 30-year career in the RCMP. A measure of his character can be glimpsed in one event during his long calling. On September 26, 1981, as a member of the Surrey Detachment Diving Team, Bob was called out to examine a fishing vessel that had collided with a freighter and was capsized, floating upside down. They did not expect it to remain at the surface much longer. Two crewmen were known to be aboard, but it was unknown if either was alive. An exploratory dive proved that only one diver could enter the vessel. Bob elected to do so, leaving his dive team member on the surface, in hopes of finding the two crewmen, despite the fact that he had neither the training nor the proper equipment for this type of rescue attempt. But with the boat sinking and trained help hours away, he decided to try.

Groping his way through the vessel in visibility of but a few centimeters, Bob eventually surfaced in a small air pocket fouled with diesel fumes. There he found both crewmen, one a nonswimmer, both alive, as well as their pet dog. Bob gave them both hasty instructions in the dive gear he was wearing, and then carried them back out through the vessel to the surface, despite having one of the fishermen knocking his mask off in a brief struggle caused by panic in the grim conditions. As if that were not enough for the day, Bob then volunteered to go back with an air bucket to save the dog (this suggestion was disapproved by his superiors; however the dog was rescued later when the vessel was salvaged).

It is likely both of these men would have succumbed to hypothermia, drowning or asphyxiation had Bob not attempted his rescue effort. For this act, Bob was awarded the Cross of Valour, highest award given to civilians in Canada for most conspicuous courage in circumstances of extreme peril. Bob was the 13th recipient of only 20 that have been awarded to date and is the only member of the RCMP to have received this award.

Bob passed away after a long illness on November 15, 2004. To honor his memory, the Canadian government named one of its nine new Hero Class Canadian Coast Guard vessels the CCGS *Corporal Teather C.V.* This ship was accepted into the CCG fleet on February 7, 2013. The naming and dedication ceremony will occur sometime in the summer of 2013.

The dive community has its own methods in honoring Bob's memory. This updated and revised edition of *Encyclopedia of Underwater Investigations* is one of those. Many people have given generously of their time and expertise to make this second edition as complete and up to date as possible. Foremost among these is Blades Robinson. Blades coordinated the revision process, working with a team of talented and knowledgeable experts in the field of public safety diving.

Others assisting with this effort included Richard Sadler and Craig Nelson (autopsy section), Terry Trueblood, William Heckler, Jeff Morgan, Michael Zinszer, Craig Jenni and John Hornsby. Additional photos have been provided by Dive Rescue International, Marine Sonic Technology, Mark O'Neill of Olympic National Park, and Brian Densmore of the National Park Service. Finally, Bo Tibbetts provided a complete content edit of the entire manuscript, adding new content as improved procedures have warranted. Thank you all!

— *Jeffrey E. Bozanic*

INTRODUCTION 1

The purpose of the *Encyclopedia of Underwater Investigation* is to remove the mystique from underwater investigative procedures and to supply a clear, descriptive, step-by-step instruction manual for the professional public safety dive team.

The public safety diver (PSD) is often called upon as an investigative arm of the police, fire department, medical examiner, or coroner extended to an underwater environment. How you fill this need as a professional investigator will depend on your desire to become part of a growing profession and on how well you prepare yourself by extending your PSD skills to include investigative knowledge.

Underwater investigation of accidents, fatalities, and crime scenes has advanced rapidly in the past decade. Unfortunately, active dive teams have gained much knowledge and education only through trial and error on the job. This method of learning has often evolved into a "trial by fire," where mistakes have been made, and both the investigating agency and the dive team's professional reputation have been compromised in the eyes of the courts and the public, in some cases endangering the lives of team members as well.

While at first glance this manual may seem to be directed toward public safety dive teams, its value to any police, fire department, attorney, or medical examiner's or coroner's office will be apparent. Never before has all this information been assembled into one manual.

INFORMATION BASE

While some aspects of this manual may be included in other specialized texts (medical journals, etc.), most information presented herein is a direct result of actual underwater recovery operations. This manual is a compilation of more than 500 underwater recovery investigations. This broad information base has not only ensured a high degree of accuracy, but it also provides detailed information and facts never before published; in short, this is an accurate, "no-holds-barred," court-ready reference manual for the investigator.

The value of this manual is twofold. First, it explains both the basic and complex details of underwater investigative procedures, so proper actions may be taken and critical errors avoided. In this light, it is truly a *vade mecum* (a handbook to be carried on-site) for the investigator. Second, after evidence is recovered and a crime or accident is being reconstructed for either criminal or civil court proceedings, this manual can serve as a ready reference for presenting yourself in court as an expert witness.

The applications of procedures explained in this manual have no boundaries. The techniques and principles set forth will hold true in all countries and all bodies of water. Careful consideration has been given in the preparation of this manual to utilize only laws that apply across all boundaries: the laws of physics, physiology, and science.

Dive Rescue International Inc. has led the world for more than a decade in the field of public safety diving. This manual is merely one more tool that will assist you and your dive team to be experts in your profession.

Note: When a person referred to in the text may be male or female, the pronoun "he" will be used for convenience; this should not be construed to mean that the person is necessarily male.

In addition, throughout the remainder of this text, when we speak of "coroners" or "coroners' offices," by extension we are including medical examiners, their offices, or any similar officials and official bodies that have comparable duties.

THE INVESTIGATIVE DIVE TEAM

Historically, police departments and others have looked upon drownings as simple accidents — a tragic inconvenience to all, but the only task at hand was a simple "body recovery." Modern problems require modern solutions, and just as specialists assist in the investigation of fatal motor vehicle accidents, specialists are now called in to assist in the investigation of all water-related deaths.

These specialists are called public safety divers. Their task is no longer a simple "body recovery" but rather to act as the eyes and hands of the police and other investigators in aquatic environments. Today, these "underwater investigators" are utilized not only by police, fire departments and coroners' offices but also by insurance companies, lawyers, and other members of the legal and paralegal professions.

To understand the application of the underwater investigator's skills, we will look at some fundamentals of what team members need to know concerning the object of a search and recovery, or the specific circumstances surrounding an accident or crime scene. Perhaps it is a vehicle, perhaps a body, or maybe there is a weapon that has been "dumped." But there is more to the story. The investigator needs to know by observing certain signs and evidence what happened, when it happened, and why it happened, then put that information together into a logical whole that will be invaluable to an investigation or a legal proceeding. It is an important role, and it is an adventure.

DROWNING — WHAT REALLY HAPPENS 2

To begin to understand the subject of body recovery, it is first necessary to understand the physiological process and implications of immersion in fluid up to and following clinical death.

To best understand this process, we will dilate time and witness an actual drowning.

DATE: Sunday, August 14

TIME: 4 p.m. (The lifeguard has just left the public swimming area).

TEMPERATURE: Air 85°F (29°C)
Water 68°F (20°C)

A 26-year-old male has spent the day at the beach with his friends. He has consumed a modest quantity of beer and has eaten mostly carbohydrates.

Our subject is a poor swimmer, but the summer heat has driven him to find comfort in the cool lake. Peer pressure and the last bottle of beer he has consumed give him the desire and the courage to swim across the lake with his comrades. The distance is a short 500 feet (150 m) — an easy swim for the others.

The group of four enters the water and begins what will soon be a race to the far shore. One hundred feet (30 m) from shore, our subject begins to tire, but being left behind by his mates is not an alternative he wishes to face.

Two hundred feet (60 m) from shore, he is tiring rapidly and glances over his shoulder briefly. He is almost halfway across the lake, and his friends are leading the race by at least 50 feet (15 m). He treads water briefly, then realizes he is closer to his friends than to the shore he has left behind. He presses on, spending more and more energy to keep afloat and less energy to move forward through the water.

At 250 feet (75 m) from the shore, he is at the point of no return, the center of the lake. His friends are now well ahead of him, and his concentration is now on his tired arms and legs. All his energy is being spent just to keep his head above water. His breathing is now labored and comes in short, irregular gasps. A small splash of water enters his mouth, and he coughs. Calling for help is impossible because his concentration and all his efforts are focused on breathing.

Assuming a vertical position in the water, he flounders for only a few seconds. More water enters his throat, and he begins to cough more violently. Between coughs he struggles to keep his head above the surface. As he coughs, he expels more and more air from his lungs; his breathing becomes shallow, rapid and less efficient. His muscles accumulate lactic acid and begin to cramp. He is tired.

Unnoticed by his friends, who are now nearing the far shore, he slips quietly below the surface. Sinking slowly at first, he loses his ability to hold his breath, and an ever-increasing carbon-dioxide buildup forces his diaphragm to contract uncontrollably.

A deep inspiration of water follows. With his lungs nearly empty, he has lost 10–15 pounds (4.5-7 kg) of buoyancy, and he sinks faster. The water now in his lungs allows the carbon dioxide in his blood to leave quickly through the alveolar bed. His burning desire to breathe is now satisfied, but the blood returning from his lungs to his heart has drastically changed its chemistry. His heart will not continue to function for much longer.

As his desire to breathe seems to come back under control, drowning loses its horror, and he sinks faster, not realizing that he is quickly losing all voluntary muscle control. In this confused and somewhat dazed state of mind, he may not feel a sinus or eardrum hemorrhage, and he may not even feel his feet touch the bottom of the lake, 50 feet (15 m) below the surface. As consciousness leaves, he instinctively grasps at the bottom silt and vegetation.

In his final moments, instinct takes over, and he breathes deeply and forcefully. His breathing may continue for a short time, and large quantities of water and silt may enter his lungs. In a confused set of actions and reactions, he may vomit, then inhale some of the contents of his own stomach.

In all likelihood, because of this strong breathing reflex, water and bottom debris may be drawn into his stomach and lungs. Additionally, the forceful inspiration of water may tear many of the small alveoli deep within his lungs, allowing lake water to enter directly into the bloodstream while his heart is still beating. At the very least, this forceful inspiration of water will admit a small quantity of blood deep into the lungs.

And his friends? They have all made the swim to the far shore without him. They look back over the route they have just swum, call his name and ask each other if anyone heard him call for help. Together, they swim back across the lake, thinking, hoping he has returned to their entry point and hidden in the nearby bushes "just to scare them."

Thirty minutes pass, and the group becomes worried. They talk to several bystanders, who "didn't see or hear anything." Someone calls the police and fire department.

When a cursory search fails to locate the subject, a series of telephone calls is made. Within the hour the public safety dive team arrives at the site.

BODY RECOVERY 3

INTRODUCTION — LOCATING THE BODY

"The public safety diver cannot and should not be held responsible for making medical observations during a body recovery."

Perhaps there was never a more inaccurate statement. Surely the public safety diver must be responsible for the first and often the most important observations. In investigative cases, locating the body is just the beginning, not the conclusion of an operation. The postmortem investigation should begin the instant the body has been located. A public safety dive team that is called on to serve in the capacity of a recovery team must first understand the intricacies of drowning, submergence and death investigation. The facts and scenarios set forth in this chapter will serve to explain not only what happens but also why.

The first step in any journey is always the most important, because the first step forward determines the direction of the journey. This chapter is designed to take the public safety diver from the beginning (how to find and where to look for a submerged body) to the end (postmortem observations). Each step should be considered carefully and understood well before proceeding. Remember: You get only one chance to do it right.

"The journey of a thousand miles begins with the first step." ~ Lao-Tse

Assessing Body Weight Underwater

What does the body of an average drowning victim weigh underwater? How heavy is it? What is the relevance of this important data in the investigation?

These are questions that are often asked. Even the experienced public safety diver may be under the illusion that a submerged human body is extremely heavy. This misconception is likely nourished by weighted dummies used in dive training exercises and reinforced by the emotionally charged situation of having to swim to the surface with a real body in tow and the drag felt when a rigor-stiffened, nonstreamlined body is moved.

Over a period of five years, approximately 100 bodies were weighed underwater. The weighing device was a simple pair of spring-loaded fishing scales. These scales were calibrated with known weights and found to be accurate within 10 percent over the entire scale (0–20 pounds/0–9 kg).

The following information was obtained using the criterion that the lightest and heaviest 5 percent would not be scored. Those weights, which did not fit into the norm, were considered atypical and were not considered to be part of a fair representation.

All bodies had been submerged for less than 12 hours in a water temperature of 60°F–65°F (15.5°C–18°C) or less than 24 hours in a water temperature of 50°F–65°F (10°C–18°C) or less than 48 hours in water colder than 50°F (10°C). Gases forming in the bodies under these conditions were so slight that the overall buoyancy of the victim was not significantly affected. These also represented average recovery times.

All bodies were weighed at depths ranging from 15 to 100 feet (5 to 30 m). The following results were obtained:

Table: Drowning Victims' Sex and Size vs. Weight in Water

Sex	Surface Weight	Less Than 40 ft/12 m	40–100 ft/ 12–30 m
Male (47 Victims)	150–170 lbs 66–77 kg	9–15 lbs 4.1–6.8 kg	12–16 lbs 5.7–7.3 kg
Male (36 Victims)	180–200 lbs 82–91 kg	7–11 lbs 3.2–5 kg	9–14 lbs 4.1 6.4
Female (12 Victims)	110–140 lbs 50–64 kg	8–15 lbs 3.6–6.8 kg	10–16 lbs 4.5–7.3 kg
Children (7 Victims)	40–70 lbs 18–32 kg	4–8 lbs 1.8–3.6 kg	5–10 lbs 2.3–4.5 kg

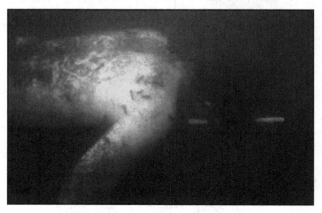

In most cases the bodies of adult drowning victims weigh between 7 and 16 pounds underwater. Clothing adds a little to their underwater weight, and they tend to resist movement by currents of less than 1.5 miles per hour.

Although these averages are taken from a limited number of drownings, the following information can be summarized:

1. Males tend to be slightly heavier than females.

2. Depth has a noticeable, but not great, effect on buoyancy. (Depths shallower than 15 feet/4.6 m were not scored.)

3. Average male drowning victims weigh 9–16 pounds (4.1–7.3 kg) underwater.

4. Average female drowning victims weigh 8–16 pounds (3.6–7.3 kg) underwater.

The most important fact available from this study is that drowning victims submerged in water are heavy enough to resist movement by water currents of less than 1.5 miles per hour (0.67 m/sec). The maximum sustained swimming speed of a diver in excellent physical condition is considered to be 1.8 miles per hour (0.8 m/sec).

Observation: If a diver can effectively swim against the current, it is very unlikely that a drowning victim will be moved. The search should always begin at the last-seen point.

Caution: Once gas formation has begun within the victim's body, buoyancy can be changed radically.

The above statistics ignore the buoyancy due to gas formation. Also, not accounted for in these statistics are drownings in less than 10 feet (3 m) of water.

Drowning victims under the age of 10 years were incredibly varied as to their underwater weight. Where great variations were noted, they were almost always on the light side of normal. Indeed, several subjects never sank. Their bodies were recovered within a few hours, floating near their last-seen points.

In the case of fatal motor-vehicle accidents in which babies less than six months of age are ejected from the vehicle into the water, it can be reasonably expected that they will float. This may be attributed

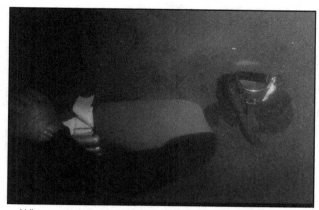

When recovering the bodies of either skin or scuba divers, additional weight is caused by compression of the foam neoprene divng suit as well as the weight belt. It is not common for a victim of a diving accident to weigh an additional 35 pounds. The breath-hold had settled nearly halfway into the silt bottom of a lake at a depth of 10 feet.

to buoyancy provided by their natural fat layer as well as air that may be trapped in tightly wrapped clothing and blankets. Disposable diapers with plastic liners contribute greatly to the buoyancy of infants.

In addition to their fat layer and the air trapped in their clothing, the breath-hold reflex of a newborn is well established. This, combined with the other factors, could well account for the fact that even when submerged, unless trapped underwater, infants usually float.

On the Scene: Fatal Motor Vehicle Accident
A motor vehicle with three adults ages 24, 26, and 45 years, two youths ages 18 and 19 years, and a 3-month-old infant left the road and plunged down a 50-foot (15-m) embankment into a slow-moving river. All were ejected (or escaped). Only one adult had injuries that could be considered fatal or near-fatal. All the others involved in the accident sustained few or no serious injuries, yet there was only one survivor.

The river was approximately 25 feet (8 m) deep, and the occupants were (except for the lone survivor) found beside the vehicle within a radius of 20 feet (6 m). The current was measured at approximately one mile per hour.

The baby was missing from the underwater scene and located, floating face-up, approximately one-half mile (800 m) downstream the same day.

The public safety dive team that conducted this mission had done their predive interviews and preparation.

Only four members were used for the recovery of the vehicle and its occupants. The remaining two divers began an immediate search downriver wearing only wetsuits, masks, fins, and snorkels. They located the infant snagged (floating) by a small log that protruded from the shore. Not only had their planning resulted in the efficient recovery of all occupants, but the underwater search was shortened. Unfortunately, the infant had died from hypothermia. No one had believed the public safety divers who suggested that a shoreline search be conducted, and they were without help.

"Whenever you reduce the time your divers spend underwater, you increase the safety of your mission."
~ Steven J. Linton

LAST-SEEN POINT
Due to the weight of a body underwater, it will settle in a near-vertical drop to the bottom.

It should never be assumed that the body of a victim has drifted away from the last-seen point until that area has been searched and found conclusively to be negative.

Fact: A body, during its descent in still water, will drift no more than 1 foot (30 cm) horizontally for every 1 foot (30 cm) it descends vertically. When this rule seems to break down, poor witness interviewing or poor witness recollection is usually to blame. It is very helpful that an investigative dive team have access to first-hand interviews with witnesses. Second-hand information, no matter how well intentioned, even from a law enforcement officer, is often inadequate for the investigative dive team. By having a senior dive team member conducting the interview, he can use his years of in-water experience, coupled with his investigative training, and draw out the relevant information for a productive underwater operation.

Binoculars — An Alternate Search Tool
An integral part of the basic search equipment of any public safety dive team should be a pair of binoculars. Where there is any possibility that the victim(s) may be on the surface, a thorough surface search, including the distant shore and local shoreline vegetation, should first be searched prior to beginning an underwater operation. While research has revealed the average underwater weight of adults is approximately 7 to 16 pounds (3.2—7.3 kg), there will always be exceptions.

Thermal Imaging — Helpful or Not?

Because of the density of water, thermal imaging technology has no benefit in locating submerged victims. If the victim is on the surface, however, and the body has not reached ambient temperature, then the use of a thermal imaging device can be helpful.

Thermoclines — Fact or Fiction?

As has already been shown, an average drowning victim, which may weigh between 7 and 16 pounds (3.2–7.3 kg) underwater, is not likely to be suspended on a thermocline (a temperature gradient in water). Such a phenomenon would occur only if the buoyancy of a human body were neutral to within a fraction of an ounce/gram.

As a body sinks, air spaces within are compressed, and it becomes heavier, or more dense, as depth increases. This phenomenon, explained by Boyle's law, demonstrates that as we descend through the water column, water pressure increases, thus decreasing the gas volume. Most divers who have worn closed-cell foam neoprene wetsuits are familiar with this fact. As a diver descends, the air spaces within the cells of the neoprene are compressed. The diver becomes heavier and must compensate for the added weight or risk becoming very heavy at depth. While the buoyancy change of a human body is not as radical, it is measurable.

Similarly, during the refloat of a body, it will ascend to the surface with its contained gas pockets expanding as it rises. This increasing buoyancy prevents a body from becoming trapped below any thermoclines that may be present.

Public safety divers have seen, on rare occasions, bodies that were about to commence refloat. In these rare instances, the buoyancy of the victim was nearly perfectly balanced at neutral. When moved, the victim seemed to hover within a few feet of the bottom, neither ascending nor descending. Once the body was raised 10 feet (3 m) or more, however, it began to rise on its own. It is believed that these bodies were only a few minutes or hours from refloat. It is this illusion that has given rise to the tale that drowning victims will be suspended on a thermocline.

When this site is carefully viewed through binoculars, one solitary hand is easily seen. Searching with binoculars is a painstakingly slow procedure but a valuable technique for the "fully functioning" dive team.

The victim of a whitewater kayaking accident was transported several miles downstream by the river current. Several teams of land-based search parties failed to find the body, which was obscured by the log and the fast-flowing water. As viewed from shore, the body is barely visible (trapped near the end of a fallen tree).

Fact: Divers, during their descent or ascent, do not become suspended on thermoclines — neither does a human body.

DAMS

Dams offer many challenging components to a dive team. Safety is a primary concern when searching these environments. Only PSD teams specifically trained in dam searches should be considered for this type of rescue/recovery effort. The most notable characteristics of dam searchers are the currents that take place throughout the water column of the dam. Downstream of the dam you may experience a downstream surface flow, and directly a foot (30 cm) beneath that you may experience an upstream flow. Within the water column (3–5 ft/1–1.5 m), the flow may be pushing downstream, whereas conditions on the bottom are generally peaceful. This is typically where you will locate the body, up until the time the body starts to incur decomposition and gas onset. When this occurs, the body will gravitate upward and begin to travel with the currents.

On the Scene: Listen to Your Witness

Two sport divers entered the water for a routine pleasure dive. The visibility was excellent below a depth of 30 feet (9 m), but the area between the surface and 30 feet (9 m) was subject to a plankton bloom, which reduced the visibility to less than 3 feet (1 m).

After a routine and uneventful dive to depths of approximately 100 feet (30 m), only one diver surfaced. The survivor stated that he became separated from his partner during a routine midwater ascent through the plankton bloom.

The surviving diver swam approximately 20 feet (6 m) to shore and waited for approximately 15 minutes. He then went back in the water in an attempt to locate his friend. When his air supply was depleted, he summoned help. A local dive club in the area began a search utilizing 15 divers. A strong tidal current was running, and the search area was concentrated down-current from the reported last-seen point. In addition to the local diving club, a team of military divers responded, but all searching met with negative results.

By sunset, 10 hours had passed, and it was concluded that the tide had taken the missing diver's body out to sea.

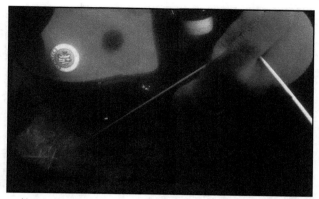

Always listen to your witness. The survivor of a scuba fatality was asked to drop a marker where he lost contact with his diving partner and surfaced. The line from the marker buoy actually passed through the hand of the missing (deceased) diver.

A public safety dive team was called in the following day. The survivor was interviewed, and, using a small inflatable boat, he directed the boat operator to the point where he had lost contact with his buddy and surfaced. At that point, he threw a small marker buoy into the water.

When the public safety dive team descended the buoy-line to the bottom, a depth of 25 feet (7.5 m), it was found that the weight from the marker was resting within 1 foot (30 cm) of the victim's body, and the white line from the marker buoy had actually passed through the victim's gloved fingers.

- Always conduct witness interviews.

- Interview witnesses separately, one witness per interviewer.

- Ask each witness to show you, rather than simply tell you, the details of their account.

- Always begin searching at the last-seen point.

CURRENT — HOW IT AFFECTS A SUBMERGED BODY

It is indeed surprising to recover the body of a drowning victim, apparently unmoved, in a turbulent river current. This phenomenon is generally a result of weight, entrapment, and/or circulatory currents (such as those that occur below waterfalls). While operating in currents is challenging, certain calculations can provide recovery divers with information that generally improves the likelihood of locating the body.

When a drowning occurs in whitewater, areas of potential entrapment should be explored. The victim's body is marked by a white surveyor's tape arrow placed on the log. The body, which had been trapped beneath the log, was barely visible from shore with the naked eye.

In rivers where the bottom configuration (large rocks, etc.) is liable to trap a body, careful searching often proves successful. This victim fell from a cliff — a height of approximately 240 feet — into a river. Her body was found 200 feet downstream from the point of impact with the water. The current was measured at 6–8 miles per hour, and water depth was 15 feet. Her body had become lodged in large rocks that covered the bottom.

After drowning in a whitewater kayak accident, the victim's body was located approximately six miles downstream. It had been totally stripped of clothing and jewelry by the current. Clothing is often removed after a body becomes trapped and will often be found downstream from that location.

This body of a youth was located in a deep pool directly below a waterfall. The force from the waterfall was not great enough to move the body in the 15-foot-deep pool. The force of the fall into the water partially removed his white T-shirt and sweater. The pool had been unsuccessfully dragged using iron hooks prior to the arrival of the public safety dive team. The victim's body could not be seen from the surface because of the whitewater outflow from the falls.

We know that in a still body of water a victim will alter its initial position above the bottom no more than 1 foot (30 cm) horizontally for every 1 foot (30 cm) it descends vertically, because radius is equal to the depth. For example, in still water of 20 feet (6 m) depth, the victim should be found within 20 feet (6 m) radius or 40 feet (12 m) diameter. Working with current involves additional calculations.

Calculate the surface speed of water by time over distance. To do this, take an object (for example, a stick) and time how long it takes to float a certain distance. Walk off 100 feet (30 m), time how long it takes to travel that distance. Let's assume it took one minute. The item traveled 100 feet per minute (30 m/minute). If we divide by 60 seconds we obtain a rate of 1.67 feet per second (50 cm/second).

If the stick traveled 100 feet (30 m) in only 30 seconds it would be traveling at 3.33 feet per second (100 cm/second). We obtain this by dividing the distance traveled of 100 feet (30 m) by the time of 30 seconds. If the stick traveled 50 feet (15 m) in 15 seconds, the rate is also 3.33 feet per second (100 cm/second).

DROP RATES OF AN AVERAGE SUBMERGED BODY

The average human body has a descent rate of approximately 1.5 feet per second (45 cm/second) in saltwater to 2.5 feet per second (75 cm/second) for fresh water. However, keep in mind there are other factors that affect this calculation, which you will learn about in upcoming chapters (Infants, Obese, Clothing). For this discussion, we will use 2 feet per second (60 cm/second) as a general descent rate. Example: 20 feet (6 m) depth will take 10 seconds to reach the bottom.

Calculation

Now let's combine the two variables into an example involving a child who slipped into a river and sank immediately. The descent rate is 2 feet per second (60 cm/second) using our average body drop rate. If the depth is 40 feet (12 m) below the bank, the time for the body to reach the bottom is 20 seconds. The public safety dive team measures the surface speed of the water as 100 feet per minute (30 m/minute), or 1.67 feet per second (50 cm/second). By multiplying the drop time of 20 seconds by the current rate of 1.67 feet per second (50 cm/second), we can

expect to find the victim approximately 33 feet (10 m) downstream of the victim's entry point into the water.

Weight

Some drowning victims are just plain heavy. A large muscle mass coupled with a thin fat layer often contributes greatly to the weight of an adult male. This, coupled with a streamlined position, could easily explain why many victims are recovered at the last-seen point, even in rivers.

Entrapment

Often natural and man-made obstructions are scattered over a river bottom. Such obstructions may include large boulders, metal cable and wires, discarded vehicles, scuttled boats, and even construction debris. These articles and others (including deep pools) may form natural traps for drowning victims.

Circulatory Currents and Waterfalls

Few people other than experienced public safety divers realize just how still a deep river pool (even under a waterfall) can be. Even when the surface water is flowing at an impressive speed, deep pools often afford a quiet, almost still area where fish, debris, and bodies may collect. When the victim was last seen near such a waterfall or deep pool, their natural negative buoyancy (underwater weight) may hold them securely beneath the white water, which may extend only a few feet deep.

On the Scene: Searching Beneath a Waterfall

A youth was rock climbing with a group when he lost his grip and fell 80 feet (24 m) into a canyon waterfall. The following day when the public safety dive team arrived, the local (land) search team suggested that the divers begin their search a quarter-mile downstream where the river became shallow, wide, and slow. Their rationale was simple: The body could not have remained beneath such a torrent of white water and must have been flushed downstream. To support their assumptions, they had already thrown grappling hooks into the pool at the base of the falls for several hours. Although they had lost one of their hooks, they had "proven" that the body had been carried away from the falls by the current. The white water on the surface prevented any of the searchers from seeing the bottom of the 15-foot (4.6-m) pool, even though the water was clear.

The use of aluminium prodding poles is an advantage when working in shallow water where current or visibility prohibits the effective use of divers. The poles should be gently tapped on the bottom. Great care should be taken to avoid damage to the body.

After two youths went missing while swimming in a canal, a search team set out with a prodding pole and effectively located the bodies during a line search.

The divers reasoned differently. It seemed possible that a waterfall of that magnitude may have cut out a large bowl beneath its base. Although they had never dived that site before, the public safety divers concluded that since it was now the summer "dry" season, the falls would be, if anything, slower and lighter than usual. It might be possible to penetrate the pool beneath the white water.

At the scene, the lone diver secured his lifeline, carefully descended beneath the white water, and swam into a deep, clear pool that was cut from solid granite. He signaled his tender that he was returning to the surface after only three minutes on the bottom.

The second descent into the pool was with a camera, and the entire scene was documented on film. When the diver returned to the surface, he carried the body of the young boy with him. The grapple hook was left behind, still firmly snagged between two large rocks.

Searching White Water

When a drowning occurs in an area of white water, the body will probably continue to move with the current until it becomes lodged. Unfortunately, a whitewater river appears as a confusion of colors, shades, and shapes; when searched from shore, the casual observer will often not be able to identify a body even when it is not fully submerged. This, compounded by the fact that fast-flowing rivers often remove clothing (and jewelry) from drowning victims when they are held stationary against the current, tends to camouflage their presence. Often land-based search teams will concentrate their search for a victim on a single highly visible item (e.g., a red jacket), not realizing that the clothing may be removed and that flesh tones tend to blend in with white water.

Binoculars and patience are often the only successful tools in searching shallow whitewater rivers. A careful, painstakingly slow scan of the river is necessary, pausing to study any irregularities in the color, shade, or shape of the surface. Each area should be carefully studied from at least two vantage points before moving on to search further downstream. In addition, sharp bends or curves in the river and natural entanglements/entrapments should be carefully searched using a prodding pole.

Prodding poles are long poles (pipes) usually made from aluminum. They should be a minimum of 7 feet (2 m) in length and light enough to be carried easily. Using a metal prodding pole allows the searcher to gently feel the bottom of a shallow whitewater river, in and around most natural entrapments. Since most whitewater rivers have rocky or coarse gravel bottoms, a metal pole is used to gently tap the bottom and areas of entrapment.

Generally, if a length of aluminum pipe is utilized, a characteristic "tink" sound is heard or felt when rocks and coarse gravel come into contact with the submerged end of the pipe. A submerged log, rubber tire or body will give its own characteristic dull "phug," which should be investigated further. Prodding poles should be used carefully and gently to avoid damage to the body.

Depending on visibility, attaching a small remote camera to a prodding pole can be beneficial, especially in calm pools in and around the search area. Placing the camera about 6 inches (15 cm) above the end of the pole gives a good view and protects the camera from damage. Ensure that the pole is marked showing the direction the camera is facing for effective searching. Remember the camera may see more than the naked eye, and video-enhancement technology can sharpen color contrasts to the point that the water appears clear.

Body buoyancy characteristics change in white water. White water is composed of 40-60 percent air. This aeration gives water a white appearance and diminishes the buoyancy provided by water by reducing the density of the liquid. Any object is buoyed by the weight of the liquid it displaces. When that liquid is less dense, it weighs less and thus provides less buoyancy. As a result, a body immersed in white water will be heavier than it would be in calm water. It will be more likely to sink and will sink at a faster rate. Even a body in which decomposition gases have begun to accumulate may remain submerged for a longer time if it is immersed in white water.

EDDIES

River eddies exist along the banks, in the middle and on the sides of moving waterways. Specific to rivers, they can pose additional problems for dive teams. An eddy occurs where fast water meets slower water that is caused by an obstruction either in or out of the water. As a public safety dive team, you should be aware that they exist and how important to the investigation they may become. An eddy may range from 10 feet (3 m) to a few hundred feet (~60 m) in length or diameter. Because they make a circular pattern in the water, it is not uncommon to find a vehicle in such an eddy, even if that point is upstream of the last seen tire tracks indicating where the vehicle entered the water. Vehicles do not immediately sink (they tend to float for a few minutes); despite their significant weight, vehicles are often carried some distance by the force of an eddy.

Last Known Activity

Nearly as important as the last-seen point is what is referred to as the last known activity.

In cases involving unwitnessed or poorly witnessed drowning, the last known activity may be the only clue to the location of the body.

Understanding the fact that a body will generally sink directly to the bottom, we are left with the very basic question: Where was the victim when the actual drowning occurred? The following sections are but a few facts that will help in locating the victim of an unwitnessed drowning.

Night Drowning

When an individual is disoriented and the shoreline is not clearly defined, he will tend to swim in the direction of light (campfires, etc.) or voices. This last known activity is simply "swimming in the direction of perceived help."

Swimming/Diving Platforms

Where a float or diving platform is anchored (a common sight at public swimming beaches), a search for an unwitnessed drowning should be conducted between the normal entry point and the float. Hundreds of poor swimmers drown each year in the United States and Canada thinking they can swim the distance. In such cases, their last known activity may have been trying to reach the swimming platform or attempting to reach the shore on their return swim. This type of behavior is often seen concerning younger males who overestimate their swimming skills or want to prove their manhood by trying feats of endurance and associated peer games. Evaluate the victim and his mental state during the last known activity.

Shallow Rivers

Every year dozens of people drown in rivers that are less than 3 feet (1 m) deep. Usually the victim is a youth attempting to cross the river on foot or an adult fisherman who slips on a rock and quickly panics as his waders fill with cold water. Either accident usually results in the body being trapped in the swift current or moved downstream a distance until it reaches a deep pool or natural entrapment.

Many such drownings involve a nonswimmer (usually a youth) who has been reported missing. When this occurs, search shallow areas first, then investigate deeper areas immediately downstream. If the missing individual was a good swimmer, immediately investigate the deeper (and calmer) pools of the river. Along with this type of search, the cliff diver who fails to comprehend the depth of the river should also be considered. In cases such as these, their last known activity may have been crossing the river, swimming in a deep pool, or diving from a rocky ledge.

The public safety diver should always be observant when searching for a missing person. Articles they may have been carrying are often excellent clues and when located will provide a last-seen point. Articles that should be carefully looked for are fishing equipment, drink containers, firearms, or even eyeglasses. The victim's last known activity may have been dropping his belongings.

Snorkeling Drowning

Each year many youngsters drown while snorkeling. It is indeed unfortunate that they are given a mask, snorkel, and a pair of swim fins, which are all too often looked upon as swimming aids. Children's watermanship skills are often poor, and frequently they have received no formal training. An unwitnessed drowning involving a youth who went snorkeling should take into account their last known activity: snorkeling.

Generally, a poor swimmer who is snorkeling will confine his activities to areas and depths where he can see the bottom without having to breath-hold dive. The search for an unwitnessed drowning in these cases should first be conducted up to a depth where (from the surface) the bottom begins to become vague. This is the very area where the victim may attempt a breath-hold dive, a very dangerous activity for a poor swimmer.

Most searches of this type will quickly meet with success if a surface search is first conducted. Snorkel drownings usually occur in relatively shallow water. This is an important consideration when responding in the rescue mode, as it lends itself to a very fast and efficient recovery. A search for a diver who is an expert breath-hold diver, however, may have to

After making their way to the vertical rock face, the survivors of the boating accident eventually swam and climbed to the point where a gently sloping rock (on the left) allowed them to climb out of the water. The sole victim's body was found directly beneath the vertical rock face in approximately 10 feet of water.

be conducted using standard search techniques, since many experienced breath-hold divers are quite comfortable conducting their activities in depths of up to 50 feet (15 m) or deeper.

On the Scene: Searching for a Snorkeler

A 9-year-old boy who was at best a poor swimmer was snorkeling in a small lake adjacent to a public picnic site. A friend who was watching from shore saw him dive beneath the surface and not return. Grief-stricken, the 10-year-old friend ran to summon the parents, who were eating lunch a short distance away.

With emotions running high, the last-seen point became more and more vague. A group of sport divers who were preparing to dive nearby began to search the area within five minutes. Their search was negative and was terminated when their last air cylinder ran dry.

The following day a public safety dive team was called and arrived on the scene. Two divers entered the water and began a surface search after speaking to the 10-year-old witness.

The young boy's body was located in 12 feet (3.7 m) of water approximately 20 feet (6 m) from shore in less than three minutes.

When later questioned, the sport divers who had conducted the primary search stated that they had searched only the deeper area directly adjacent to the

Several sport divers had failed to locate the victim of the boating accident even though it was plainly visible from the surface. Adequate witness interviewing and accident reconstruction is a necessary part of predive planning. The body of this victim was located by the public safety dive team prior to putting any divers in the water.

shallow shoal jutting out from the shore. Interestingly enough, their search had taken them to depths of nearly 100 feet (30 m). To reach the drop-off they would have had to swim directly over the young boy's body. They had failed to carefully search the shallow area in their haste to reach deeper water. In this case, the last known activity was snorkeling — in shallow water.

The Nearest Shore
When faced with cold water and a likelihood of drowning, the victim often recognizes only the need to escape the water, not the ability to do so.

Swimmers will often swim to the nearest land. This would-be exit point (which may be the nearest) may not be the easiest. In areas where the nearest shore is a steep cliff, the primary search area should include the bottom directly below this rock face. The nearest land is referred to as the point of attraction. It is the visual reference that may stand out to the swimmer in trouble merely because it is the most obvious. Many points of attraction become traps.

Points of attraction include steep cliffs that enter the water vertically, a fallen tree sloping out from shore, or a log boom. Fallen trees especially are efficient traps since a tired swimmer, once reaching the outermost end of the tree, may have no strength left to negotiate the tangle of branches necessary to progress down the tree to shore.

Swimmers are often attracted to these points of attraction, and, having reached them, do not realize until it is too late that they will have to swim even further to escape the water.

When conducting a search for an unwitnessed drowning victim, if an area can be determined where the victim entered the water (last-seen point, upset boat, fishing tackle on the bottom, etc.), consider searching in the direction of the nearest point of attraction.

On the Scene: Boating Accident
A small boat capsized in the Pacific Ocean on New Year's Day. The waves were 3–4 feet (1–1.3 m) in height, and the accident occurred approximately 100 feet (30 m) from shore.

All four victims swam to shore. In this case the shore was a steep, rocky cliff covered in barnacles. Waves, cold water, and an impossible exit resulted in an "every-man-for-himself" survival swim.

Three survivors exited the water approximately 100 yards (90 m) from their first contact with the cliff. A flat rock jutted out into the water and allowed a difficult but possible exit.

Hypothermia had all but erased their collective memories, and, as witnesses, they were unable to even describe where they exited the water. A general shoreline length of approximately 500 yards (460 m) was suggested by one of the survivors. Their memories were further clouded by the fact that after exiting the water, they had to walk several miles to the nearest house.

The following day a local diving club (seven divers) searched the designated area with negative results. As an afterthought, the local public safety dive team was called in. They carefully interviewed the witnesses and noted that the palms and fingers of all survivors were badly lacerated from barnacles. One survivor had required medical attention and showed up for the interview with both her hands bandaged. This fact alone was a clue that much time had been spent in the water in an area where there was a cliff and barnacles.

Diving under a log boom requires special consideration and training. The diver should be well controlled by his tender and never allowed to surface within the confines of the log boom. Wave action caused by a passing boat can easily cause the logs to brush together, crushing the diver.

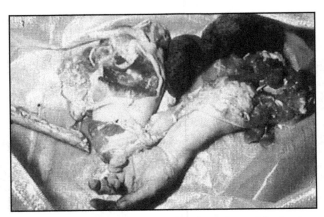

A man jumped from the stern of a large ferry while it was unloading passengers and vehicles. His entry into the water was witnessed, but no body was found. A careful search by public safety divers proved successful. The most intact portion of the body was a hand. Subsequent fingerprinting proved identity. Other remnants were located and retrieved by the divers, who conducted a two-day search to retrieve everything possible. Recover all evidence.

The dive team swam along the surface of the water next to the barnacle-covered cliff. The water was only 10 feet (3 m) deep, and the bottom could be seen from the surface. Thirty feet (9 m) out from the cliff, the bottom fell away sharply. The body of the drowned victim was seen by the divers (from the surface) against the cliff, resting in only 10 feet (3 m) of water.

No one had searched next to the point of attraction. In this case the victim's last known activity was trying to exit the water with his friends.

Industrial Accidents — Log Booms
Each year the lives of numerous workers are lost merely because of their failure to wear life-saving devices. A common drowning of this sort involves those who work on and around log booms.

The standard dress for these workers is warm (often bulky) clothing, a hard hat, and heavy leather "cork" boots. These cork boots are heavy leather work boots with soles implanted with metal spikes. They are designed to provide a good grip for the worker as he jumps from log to log.

Many drownings of this type are unwitnessed. Often a cry for help is not even heard. Either it was never made, or other boom boats working in the area covered up the victim's cries for help. Log booms are usually secured in booming grounds,

areas designated for this purpose. Since tree bark is constantly being stripped from these boom logs due to their continual bumping and rubbing against each other, the bottom of the lake or river in these areas is usually a wasteland, covered in a foot or more of bark mulch in various stages of decay. This is easily stirred up by boats and divers and makes searching very difficult indeed.

When searching for a victim of a log-boom drowning, the victim will nearly always be located under the outer edge of the boom if the boom is full. The reason for this is simple: When a worker falls in the water within the confines of a log boom, he is surrounded by smaller-sized logs that are usually packed fairly tightly. The worker will have many handholds within easy reach on all sides, which he can take hold of to raise himself out of the water.

When a worker falls from one of the larger logs (chained end to end) that constitute a perimeter containment, he must climb out of the water over one rolling, large, slippery log. Wearing cork boots and bulky water-soaked clothing does not make this an easily-accomplished task. This person's only alternative is usually to swim (often holding on to the larger logs from time to time) along the perimeter of the boom until he reaches the shore. If hypothermia and fatigue win in the struggle to reach shore, he will eventually slip beneath the surface and drown directly below the outer edge of the boom.

There are two exceptions to this rule: The first (perceived) exception is when the boom is allowed to drift into a new position due to wind, tides, or other currents. The other exception occurs when an individual falls into the water inside the boom, and a passing deep-hulled boat leaves a wake that agitates all the logs within the boom. Few commercial logs weigh less than 500 pounds (225 kg). Being caught between several of these logs, banging together, usually results in severe, often fatal, crushing injuries.

Diving under or around log booms requires special consideration and training. It is a very hazardous activity that demands expertise and discipline. Searching under and around log booms is even more dangerous than diving beneath an ice canopy.

Drowning from Large Vessels

If a victim should fall or jump from a large vessel and its propeller is turning at the time of the accident or suicide, then public safety dive teams may be presented with a unique problem.

Many large vessels, most commonly ferries that carry vehicles, maintain an aggressive forward thrust when loading and unloading passengers and vehicles. This positive thrust ensures that the vessel will be held tightly to the docking berth during the cargo transfer.

People who enter the water — either in an act of suicide or by accident — may be quickly swept through the propeller and disappear from sight. When this occurs, the public safety dive team may not be looking for a body.

Depending on the size and speed of the propeller, the victim's body may be badly dismembered and will have to be recovered piecemeal. It is not uncommon for the largest portion of the body to be a mere few pounds/kilograms in weight. In cases such as this, a painstakingly slow and methodical search must be carried out. Hands, feet, individual bone fragments, unknown tissue and organs, etc., must all be recovered carefully. Even the heart, if recovered, may lead to an autopsy finding to prove the fall into the water was due to a heart attack. In cases such as this, death would be declared not suicidal (or homicidal); not only could the investigation be concluded, but the victim's family might be paid the full value of a life insurance policy whose terms excluded payment for death by suicide.

In particular, hands (for fingerprints), the head, and jaw fragments (for dental examination) are crucial for victim identification. A serious attempt should be made to recover all body parts. (Refer to Chapters 12 and 13.)

Caution: When conducting searches around any vessel, take great care to ensure that the vessel's engines are not running and cannot be started while divers are in the water. It is important to utilize a "lock-out, tag-out" system to prevent accidental activation of equipment aboard the vessel that may kill or injure dive team members.

Suicidal Drowning

Although suicidal drownings are not common, public safety dive teams will inevitably be called in to search for and recover the body of a suspected suicide victim. In cases such as this, it cannot be overstressed that all suicides, indeed all deaths, should be treated (and investigated) as homicides until all possibility of foul play has been ruled out.

Usually in suicides a note will be left behind by the victim, and it will indicate the approximate location of the body.

An individual who plans and then carries out a suicide by drowning has usually planned a deliberate course of action. Such people, while they may be in a very melancholy state of mind, are often thoughtful and considerate. In many cases, the death is carefully planned so as to cause the least inconvenience to others. Often (if not indicated in a suicide note) their chosen sites will be below a high cliff, an open area of a partially frozen lake, or the deep outer end of a nearby wharf. Routinely, clothes will be found at the point of entry, often neatly folded and carefully stacked. On the top of the pile is often a wallet, purse, identification, a suicide note or a possession the decedent considered valuable and would like to see passed on to a friend or relative. This collection or pile of belongings is often referred to as the "suicide headstone" — it marks the grave. The public safety dive team should read carefully suicide notes left behind; they may contain valuable clues as to the location of the body or other evidence.

In cases involving a suicidal drowning from a wharf, the choice is often made to weigh themselves down with scraps of iron, chain, lead weight, concrete blocks, or anything conveniently available. It is also a common practice for suicidal individuals to make an attempt at tying or binding their own wrists and/ or ankles and legs to ensure a sure and swift outcome of their planned action.

Another characteristic of a wharf suicide is the practice of tying a light line to either his own body or the weight to which he has secured himself. The opposite end of this line is then secured to the wharf or a piling. This is merely an attempt on his part to ensure that his body is located and recovered as quickly as possible.

It is indeed sad to think that a person, any person, in such a state of mind would choose a careful, deliberate plan that would show so much consideration for those left behind.

Suicide notes often reflect their authors' concern that their bodies be found so that life insurance policies may be paid to their spouses and families. In contrast, some suicide victims are trying to send a message to their family members, attempting to cause family members guilt and highlighting disagreements they may have had. Typically a note or investigation will more fully explain the victim's position, effectively giving them "the last word."

Caution: When recovering the body of a suicide victim, it is imperative that all knots remain undisturbed. If possible, the body should be photographed in the position it was first discovered. An alternative to this would be raising the body and allowing the knots and ligatures to be inspected by a trained investigator on the surface. Only as a last resort should the weights be removed prior to recovering the body. If bindings, ropes, etc., must be removed, they should be cut where there is no knot present. Once on the surface, the knots should be secured. This can be done by passing a needle and thread through the knot several times. The knot may be crucial evidence.

It would be unfortunate to later discover that the suicide victim was tied to the weights with a knot that was only remembered to be a bowline, only to find out that the victim never had any rope skills and could not have possibly tied such a knot. The knot is evidence; it must be preserved.

BATHTUB, HOT TUB AND POOL DROWNINGS

Even though a dive team is not always called to the scene of a bathtub drowning, this is a very important segment within underwater crime scene investigations that warrants attention from law enforcement and emergency response personnel. Careful measures must be taken to preserve evidence and investigate the drowning.

INVESTIGATION OF DEATH 4

OVERVIEW — HOMICIDE OR ACCIDENT?

Whether the death of an individual was the result of an accident or a homicide, only a trained observer/investigator should be utilized in the recovery of the body.

Beginning moments before the cessation of a viable heartbeat, the body becomes a clock. As each minute, hour, day, week, and month passes, the human body, in death, becomes a court-ready exhibit — a "book," which if read carefully will supply critical information, answering questions that in the past have remained a mystery.

An investigation of death must answer the following questions: who, when, where, and by what means death was met. Failure to completely answer these questions will result in a weak link in the chain of evidence. This weak link can allow fatal accidents to be repeated (accidental deaths), families of victims to go uncompensated for the loss of a wage-earner or loved one (civil proceedings/court action where death was due to negligence), or a guilty person to remain free (homicide).

To sufficiently explain the details leading up to and surrounding death, the investigator should be able to read the body and understand the information it presents.

To do this effectively, the investigator must possess a basic understanding of a science referred to as postmortem physiology, the study of the changes that occur in the human body at the time of and following clinical death. These changes may continue for months or even years.

Since these observations are transient in nature, it is of great importance that the trained investigator be able to observe and record details that may later be lost to the pathologist during the postmortem procedure commonly referred to as the autopsy. Since there are many changes that take place in the human body following death, and these changes continue for days, weeks, or even years, it is of paramount importance that the investigative public safety dive team possess a basic understanding of postmortem physiology so they can relate their findings clearly and accurately to the police, pathologist, and coroner. Their findings may not be the same as those noted by the pathologist, but both sets of observations, when taken in concert, may relate a story.

The chronology of events, the order in which postmortem changes occur, and the factors involving the speed at which they proceed is of great importance when making and recording observations. Since several events may have taken place simultaneously, the order cannot be strictly adhered to in this text. Instead, each postmortem artifact will be discussed separately.

Before beginning a study of postmortem physiology, it should be understood that all postmortem artifacts are subject to many variables. Understanding how these variables affect the rate of change of a human body (deceased) is even more important than merely being able to identify what is happening.

Possessing the knowledge that will enable you, the investigator, to make critical observations and present a detailed and meaningful report will be the first step in becoming an expert witness.

CAUSE, MANNER, AND MECHANISM OF DEATH

Coroners, medical examiners, and forensic pathologists are always seeking to find the cause, manner, and mechanism associated with any death. It is important to understand each term and how it is used with postmortem examinations.

As it relates to postmortem investigations, cause of death is the physical action that resulted in death, such as gunshot, heart attack, stabbing, or drowning. Ask yourself *what* happened to lead to the decedent's death.

The manner of death directly relates to five areas that include natural, suicide, homicide, accident, or undetermined. Ask yourself *why* did this death come to be?

The mechanism of death ties together both the cause and manner. For example, in the case of a stabbing, the stab would result in blood loss (hemorrhage). Ask yourself *how* this death occurred. Another example might be drowning (cause of death), and the mechanism would be suffocation due to anoxia. It is worth noting that during the postmortem examinations there may be several mechanisms related to cause of death.

Postmortem Physiology — What Is It?

Postmortem (after-death) physiology (the science of the function and phenomena of living things) seems to be a contradiction in terms. It is not. Death in and of itself cannot be covered by one simple definition. Indeed, after the irreversible phenomenon referred to as the death of a person occurs, many organs and tissues in a human body continue to live, if for only a few minutes or hours.

When cellular death is complete, other organisms within the body multiply and feed on the tissues. This function is the natural process of decay. The breakdown of the human body after death is merely a recycling of its vital nutrients back into the biosphere. In death, the human body becomes an ever-changing environment for a variety of biochemical and biological changes.

In effect, the human body without spirit or life of its own becomes one that is once again "alive." Because of these various life-forms, most of which are microscopic, the term *physiology* is indeed accurate.

Postmortem Physiology — Its Importance

The study of this subject encompasses not only the changes that take place in the body but also the very nature of the body itself.

For the purpose of this manual, the emphasis is on studying postmortem physiology as it relates to an aquatic environment. Many of the principles, however, are universal.

Side scan sonar can provide an image of a sunken object even in low or no visibility conditions. It is particularly useful in searching large areas of flat bottom (like lakes). Using side scan sonar might take hours, where using divers might take days or weeks to do the same thing.

A thorough working knowledge of postmortem physiology will assist the investigator in many areas, including locating the drowning victim, ascertaining the cause and time of death, and writing a complete court-ready investigative report, the hallmark of a professional.

Postmortem Physiology and Locating the Body

Whenever a body is submerged in water, whether it was an accidental drowning, suicide, or a homicide, the first stage of the investigation must be a search for the body. With the advent of sophisticated electronic equipment, successful searches for even the unwitnessed drowning victim are becoming more common. These sophisticated devices include remotely operated vehicles (ROVs), side-scan sonar, and sector-scan sonar.

ROVs are vehicles that can be guided from the surface using a joystick or other control devices. They typically have integrated video cameras, so the operator can observe the vicinity in which the ROV is in operation. Most have depth sensors, compasses, and thermometers. Some have graspers, claws, or mechanical arms that be used to recover evidence. One advantage of using an ROV is that a victim can often be located before deploying a public safety dive team, increasing efficiency and reducing risk. ROVs range in size from that of a shoe box to units weighing several tons. Operational depths vary but may be as deep as thousands of feet (meters). ROVs are becoming more sophisticated but have limitations. Public safety teams typically utilize an ROV in conjunction with other technology so personnel know the precise location of the ROV, and they can document the area that is searched. Without a rotating type sonar or another technology to document the location of the ROV, the ROV search can be haphazard and ineffective.

Side-scan sonar and sector-scan sonar are technologies that may allow a broad search area to be efficiently viewed. They work best on flat bottoms with few obstructions. Although sonar has existed for decades, not all units work well in public safety diving operations. Sonar units now exist that possess capabilities that make them useful tools in locating bodies on the bottoms of lakes and rivers.

Side-scan sonar technology allows a team to search a large area quickly. In areas where the bottom is clear of debris a trained sonar operator can search an area approximately the size of a football field (48,000 square feet / 4,459 square meters) in about one minute. While experienced sonar operators can locate submerged victims, an operator with less experience can identify targets that approximate the shape and size of victims and clear large areas so dive teams and ROV teams can concentrate their search efforts to areas with the highest probability.

It is important to note that technology does not replace a well-trained public safety dive team. The technology only helps the team be more effective.

Some public safety dive teams have attempted to utilize cadaver dogs to locate submerged victims. The rate of success varies considerably, but most will acknowledge that a cadaver dog can determine only where the scent comes to the water's surface, not where the submerged victim is located. If a team chooses to use a canine, then there should be some interaction before an actual deployment to gain confidence in the abilities of the dog and the handler. Additionally, the search dogs need to be familiarized with working aboard boats and around divers. All search dogs and handlers should provide proof of certification by the International Congress of Police Service Dogs or another recognized organization prior to being involved in a recovery operation.

Caution: The mere purchase of technology does not guarantee success. Professional training conducted with a unit suitable for this purpose is mandatory if any acceptable degree of success is to be expected. Even with these devices, however, a professionally trained public safety dive team will always be mandatory. Machines, be they sonar, video, or whatever, can only assist.

Searching Priorities

Unless the body of water is incredibly small or the victim is already marked with a buoy, a search will be necessary. The first priority in the formation of any search pattern or plan is to locate areas of high probability where the search should begin, whether by ROVs, sonar, and/or trained public safety divers.

Myths and Misconceptions

To begin understanding fully what happens to a body after a drowning (or any submergence in water), many myths and misconceptions must be cleared away and replaced by fact. The next section will detail the observable physiological changes that occur during and following drowning.

Postmortem Observations Relating to the Cause of Death

A report by the World Health Organization estimated 388,000 annual drowning deaths worldwide in 2006 and listed drowning as the third-leading cause of unintentional injury/death worldwide. Drowning deserves close scrutiny by the investigator because submerging a human body in water is sometimes used as a means of disposing of human remains and concealing a crime. In these cases the investigator must be able to differentiate between the effects of death by accidental drowning and death by homicide.

Drowning, even accidental drowning, should be investigated thoroughly. An inadequate investigator can only conduct an inadequate investigation. In any death, the evidence is always present — the absence of knowledge, however, will result in evidence not being noticed or documented.

The Physiology of Drowning

When a drowning occurs, a significant set of physiological responses occur as a precursor to clinical death. These responses usually continue for seconds or even minutes after consciousness is lost. It is to our advantage that they result in often-observable phenomena. As a public safety diver, you should understand the physiology of drowning, since you will be the first to make the critical observations necessary for a detailed investigation and complete an acceptable court-ready report.

Inhalation of Water

As a swimming or submergence scenario progresses to an actual drowning, a specific series of events takes place. Following submergence of the face in water, an attempt at breath-holding (apnea) is the first line of defense against water's entrance into the lungs. As this phase progresses, a panicky struggle quickly develops. As the body's energy reserves (chiefly oxygen) are expended, a sequence of events occurs that will culminate in clinical death unless interrupted. This chain of events is as follows.

1. Voluntary/involuntary breath-holding occurs until the concentration of carbon dioxide in the blood reaches a level where stimulation of the respiratory center in the brain takes command.

2. A small amount of water (5-20 cc) is inhaled into the trachea.

3. An involuntary coughing reflex follows. This is a natural reaction designed to expel foreign material entering the trachea. Each forceful cough expels some of the air in the lungs.

4. A greater quantity of water is inhaled into the trachea and eventually into the lungs with each attempt to expel the water present. This cycle may be interrupted by a bronchial spasm in which the epiglottis closes forcefully, forbidding further entry of water. This spasm may continue on past the loss of consciousness to the point of clinical death.

5. The increased coughing reflex, combined with water entering the bronchial tubes, may trigger a vomiting reflex wherein the stomach contents are first expelled and then aspirated into the lungs. As more and more air is expelled from the lungs, buoyancy of the body is lost, and the next phase follows.

6. Profound unconsciousness due to anoxia develops. Convulsions usually follow with more vomiting and deep agonal gasping. This unconscious gasping reflex may be so profound that

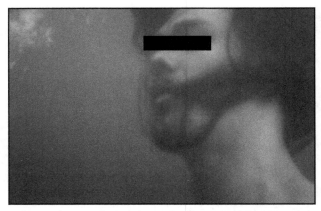

During the agonal-gasp phase of a drowning, the victim may aspirate (breathe into his lungs) water, bottom debris or even his own hair. During the autopsy of this drowning victim, hair as well as silt from the lake bottom were recovered from the lungs.

water and bottom debris, along with vomitus, are drawn deep into the lungs and back down into the stomach. Respiratory standstill follows quickly.

7. Because the lungs are now flooded with water (either saltwater or freshwater), the blood that is circulating within the lungs undergoes a radical change in its chemical balance. Cardiac arrhythmias leading to ventricular tachycardia and fibrillation ultimately precede clinical death.

8. Within two minutes following the agonal-gasp phase (in freshwater drownings), sufficient water is absorbed directly into the bloodstream to account for nearly 50 percent of circulating blood volume. This chain of events may be altered by an antemortem struggle. The greater the struggle, the shorter the cycle.

"Wet" vs. "Dry" Drowning

In recent years much has been said and many theories put forth regarding "wet" vs. "dry" drowning. Although many members of the medical profession still disagree, there are several myths and misconceptions that should at least be discussed.

Wet Drowning

A wet drowning is usually defined as a drowning where the lungs have become partially or wholly flooded with water or any other drowning medium. Most drownings are classified as wet.

During the last moments of (unconscious) life, even a profound spasm of the epiglottis will ultimately relax, allowing the airway to reopen. The terminal convulsions that follow usually result in aspiration of water deep into the bronchioles.

During the autopsy, when the lungs of the victim are removed and weighed, a noticeably heavier-than-usual weight will appear on the pathologist's scales. This is often the only measure utilized to indicate a wet drowning.

Evidence of a wet drowning will be obvious during a complete autopsy. Since the lungs will be voluminous (bulky and ballooned-looking) on removal from the body, they will retain their size and shape and possess a characteristic firmness. This alone, however, does not necessarily indicate a wet drowning as we may consider it. Deaths associated with arteriosclerotic heart disease often demonstrate severe pulmonary edema (fluid effused from the lung tissues themselves). During postmortem examination, the lungs of such a victim will often appear heavy, and an incision made directly into the lung will demonstrate large quantities of a foam similar to that seen in a wet drowning. Similar pulmonary artifacts may be seen in cases of chronic alcoholism and epilepsy. Pulmonary edema is common in deaths occurring either during an epileptic seizure or in cases of advanced chronic alcoholism.

Dry Drowning

Dry drownings have been said to occur in 10–12 percent of all medically documented (autopsied) drownings. Some references quote an incidence of dry drowning as high as 20 percent.

The dry drowning phenomenon is usually explained by describing a glottic seizure, or a tight closing of the epiglottis, during the active phase of drowning. As has already been explained, this spastic glottic musculature usually relaxes after consciousness is lost. The convulsion and agonal gasping that usually follow would seem to deny the very possibility of a dry drowning. The retention of an established laryngospasm past the convulsive stage would seem to be difficult to explain.

One possible explanation that has been accepted, however, is the phenomenon of cadaveric spasm.

During the agonal moments, any muscle or muscle group actively being utilized may display a spastic contraction that continues even after clinical death. This phenomenon is a rare but important observation during any body recovery and will be discussed at length later.

A cadaveric spasm could be responsible for what has been called dry drowning, and it would seem possible that this phenomenon could result in a purely asphyxial death as a result of submergence in water. It is suggested that any diagnosis of dry drowning following the recovery of a victim who has been subjected to aggressive mouth-to-mouth and cardio-pulmonary resuscitation would not be accurate. During this lifesaving procedure, the lungs of the drowning victim are inflated and the heart is pumped mechanically. If this is done efficiently, water inhaled by the victim into the lungs will quickly be absorbed into the circulatory system through the alveoli.

Even so, not all such (unsuccessful) rescue attempts will result in an autopsy finding of a dry drowning. Pulmonary edema may still develop, and even hours after the victim's submergence, pulmonary edema may be a primary cause of death. This, incidentally, is the so-called phenomenon of secondary drowning and is a primary reason why all successfully resuscitated victims should be given immediate medical attention and hospital admission.

Statistics indicate that dry drowning in infants is much more common than in adults. There is no definite explanation for this, but evidence indicates that the breathing reflex is not as well developed in infants and children as in adults. In youngsters especially, the agonal-gasp phase of drowning may be diminished or totally absent.

Observations

When a body is located, whether it is underwater or on land, it should be carefully studied and photographed (when possible) prior to removal. During a drowning, the following observations should be made; the recording of these observations should become an integral and routine part of the investigative report.

1. Look carefully around the head, face and mouth of the victim, as well as on the bottom, for any

signs of vomitus. This should be the first observation made, as this evidence is very transient and could easily be removed by current or your own swimming action.

2. Observe the region of and, if possible, inside the mouth. Are there any signs of the victim having inhaled silt, bottom debris, vegetation, or even his own hair?

3. Consider using plastic bags and large rubber bands to secure the head while underwater to preserve trace evidence prior to removal. Consider the benefits of bagging the hands and feet also. Be sure to note the process in your report.

4. Under certain circumstances, taking samples of fluids or solids on or about the victim (especially one enclosed in a vehicle) may prove beneficial. Using clean and inexpensive turkey basters, an investigator can collect blood, fibers, and other trace items for later evaluation in the case. This potential evidence is transient and will otherwise be washed away in most cases.

If an investigator finds vomitus around the head, face, or mouth, or if he finds inhaled silt, bottom debris, or hair, this would indicate that the victim was (likely) alive during submergence. The absence of any such evidence on cursory examination, however, is not a positive indication that the victim was not alive during submergence.

Caution: The presence of bottom debris in the mouth and trachea could be a result of currents or wave action in shallow water. The presence of vomitus is a more reliable indication that the victim was alive during submergence. Even so, an individual who died as a result of a fatal drug overdose may have vomited prior to death. On occasion, submergence in water is merely an act by another person to conceal a crime or dispose of a body.

Lung Examination (Blood)
During the agonal-gasp phase of drowning, many tears may develop in the lungs' alveoli. Normally these small air sacs are strong enough to support normal breathing activity with no damage. When a forceful attempt is made to breathe a medium as viscous as water, however, many of these alveoli become torn, allowing water to enter directly into the bloodstream. Conversely, blood may also enter the lungs. Small spot-like hemorrhages may be seen on the surface of the lungs in autopsies of drowning victims. This is a result of overdistention of the lungs by the drowning medium. These small spots may have a faded pink-red appearance. This faded appearance is due to dilution and hemolysis caused by the inhaled water. The assumption that these observations can be made only during an autopsy is wrong

Observations
The escape of blood from the mouth and nose is often seen during the delivery of the drowning victim from the bottom to the surface. The main origin of this blood is from the torn/ruptured alveoli that were damaged during the agonal-gasp phase of drowning. This blood may continue to be seen coming from the nose and mouth even after removal of the body from the water, especially if it is left in a prone (face-down) position after recovery.

Caution: Blood may also be from other sites and/or as a result of injury or blunt trauma, as in the case of a homicide, motor vehicle accident, or even a failed rescue attempt. In the absence of any obvious blunt trauma to the facial area (including the inside of the mouth), the most common origin of blood seeping from the victim's nose and mouth is the lungs. This observation would tend to indicate that death was due (in part, at least) to drowning.

Lung/Trachea Examination (Foam)
The hallmark of a drowning is often said to be the meringue-like foam that is abundantly exuded from the mouth and nostrils of the typical drowning victim.

This foam is commonly (though not always) first seen shortly after the deceased is removed from the water, and may continue to be produced for several hours. This foam has three sources:

1. During the agonal-gasp phase of the drowning and the subsequent overly vigorous struggle to survive, mucus is produced in abundance by the tracheal and bronchial glands. This mucus is secreted as a result of irritation caused by the aspirated (inhaled) water and is quickly mixed into a foamy consistency.

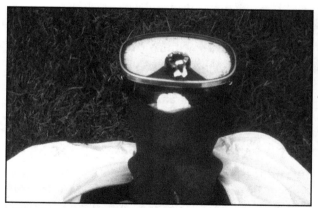

The foam that is produced as a result of drowning may completely fill the mask and be generated from the mouth and nose of the victim in a characteristic mushroomlike fashion.

2. Foam is produced deep in the lungs in response to alveolar lining cell irritation by the inhaled water. This edema fluid that is produced deep within the lungs will continue to be produced and will escape from the mouth and nostrils for several hours after removal of the body from the water.

3. During the agonal-gasp phase, as millions of microscopic alveoli are ruptured by the forceful antemortem gasp, blood plasma may enter the lungs. This plasma and its many complex chemicals will also add to the production of foam and may even add to the bloodstained appearance that is often documented in drowning victims.

Observations

Shortly after the recovery of a body from the water, large quantities of bloodstained froth may be seen issuing from the nose and mouth of the victim. This would indicate (albeit not conclusively) that water was inhaled into the trachea and lungs and would tend to indicate that the victim was alive when submerged.

Caution: This foam, although common, is not always seen in drowning victims. When it is seen, its production ranges from minimal to quite profuse. The blood content of this foam may also range from nil to predominant.

A similar foam production is often observed in individuals who have died from acute left ventricular failure with massive pulmonary edema. Unfortunately,

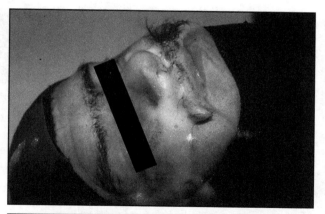

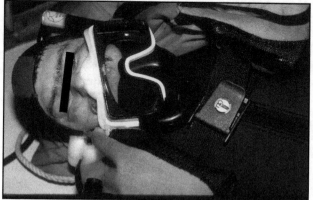

The victim of a scuba fatality recovered by public safety divers within four hours of the accident shows the characteristic foam. This foam may continue to be produced for many hours after removal from the water.

many drownings occur as a combination or sequence of events; e.g., a heart attack may have been the cause of the drowning.

POSTMORTEM CHANGES RELATING TO THE TIME OF DEATH
Definition of Death

Until recent years, the definition of death was a simple one. With advanced medical technology and case law, however, it has become more difficult to define. *Black's Law Dictionary* defines death as "the cessation of life; the ceasing to exist; defined by physicians as a total stoppage of circulation of the blood, and a cessation of the animal and vital functions consequent thereupon, such as respiration, pulsation, etc."

This definition has been utilized in subsequent court decisions. In the case of *Thomas v. Anderson*, a California District Court of Appeals said, "Death occurs precisely when life ceases and does not occur until the heart stops beating and respiration ends.

Death is not a continuous event and is an event that takes place at a precise time." This definition is a legal definition contrived for court purposes. In truth, death is a much more complicated matter.

The science of postmortem physiology must consider a greater truth. While the loss of a viable legal conscious life may be an instant event that takes place "at a precise time," the human body is composed of many different types of tissues, with each specific tissue or organ possessing the ability to survive for a different period of time in the absence of the circulation of blood.

In years past, anyone recovered from water following a submergence in excess of six minutes was automatically declared dead. Research and efforts by Dr. Martin J. Nemiroff proved that this was simply not true. Indeed, a viable life can exist for up to what is now termed "the golden hour." Clearly, a legal definition of death does not apply, since "near-drowning" victims are "legally dead" upon recovery. Nature follows a different definition. The recovery and successful resuscitation of near-drowning victims, coupled with advances in medicine in recent years, has led to a definition of cerebral (brain) death. The usual criteria to be met for a person to be declared dead include:

1. Bilateral dilation and fixation of both pupils.

2. Cessation of breathing without mechanical assistance.

3. Cessation of spontaneous cardiac action.

4. Absence of all reflexes.

5. A completely flat brain wave tracing.

Even when all these criteria are present, individual muscle fibers and cells may be kept alive for a considerable length of time. Hospitals routinely keep individual bodies alive using mechanically assisted cardiac activation and mechanical resuscitation. This activity is commonplace when the deceased has arranged to donate an organ and that organ (kidney, liver, heart, lungs, etc.) must be kept alive while the recipient is being made ready.

Death of an individual as a viable being is referred to as somatic death, while the ultimate death of all cellular elements is known as cellular death. While

somatic death may be defined as an event that "takes place at a precise time," it must be understood that cellular death within a body is a continuous process. It is this process of death and the changes that take place during and after this process that set into motion the clock and calendar of postmortem physiology.

ESTIMATING THE TIME OF DEATH
Introduction
As will be explained, the estimation and determination of the time of death is, at best, an estimation based on many observations. Estimating time of death is almost more art than science because there are numerous factors that cannot be known, making accurate estimates difficult. This is especially true in underwater investigation.

There is no one technique or test that can be accurately used to determine the time of death. In the past, many procedures have been tried, some even accepted into medical journals and courts as accurate. However, as science progresses and a greater database is made available, these so-called "truths" have all been challenged. In the final outcome, no one test has stood the test of time when utilized independently of all others.

Conversely, when several techniques or postmortem observations are employed to determine the time of death, or the postmortem interval, their findings can be used to support each other. For this reason, observing and collecting as much information as possible is essential; if an open-minded, broad approach is employed, the postmortem interval can be accurately assessed.

There is one truth, however, that should be kept in mind when attempting to determine the time of death: When determining the postmortem interval, the greater the interval, the greater the margin for error.

Eye (Ocular) Changes
The eyes of the victim are a ready, easily accessible, and useful source of information relating to the cause and time of death. This information, which is available to the investigator, is transient in nature and will not appear (as it did at the time of the body recovery) to the medical pathologist. For this reason, it is important that observations of

the eyes be made and recorded at the time the body is recovered.

The areas of the eye that must be observed are the cornea (the clear, raised portion of the eyeball covering the pupil and iris) and the sclera (the white, opaque portion of the eyeball).

If the eyelids remain open after death and the eyes are exposed to air, changes that are caused by drying will be evident. If the ocular globe (eyeball) is exposed to air after death, a visible film forms. This film covers the cornea and the visible portion of the sclera and may be visible within the first few minutes after clinical death. Conversely, if the eyelids remain closed after death, or the victim is submerged in water at the time of or immediately following death, the eyes may retain a lifelike, glistening appearance for several hours.

In addition to this film, other changes take place that could be important indicators as to the time and cause of death. A noticeable clouding of the cornea appears 12 to 14 hours after death. This cloudiness is unlike the thin "drying" film, which may appear within a few minutes; it gives the cornea a frosted-glass appearance when it is fully established. Its appearance has been documented in less than two hours, but it usually appears in 12 to 24 hours. Complete corneal opacity (clouding) can be expected in three days under ordinary conditions. Submergence in water will delay this process, however.

Along with the corneal changes, the sclera will become discolored with time and lose its healthy white appearance at a rate similar to the time it takes for corneal changes to take place. This discoloration is known as *tache noire sclerotique*, and it is the result of the eye being exposed to air. In most deaths, the eyelids remain at least partially open. It is because of this fact that observation of the condition of the eyes is critical, because in most cases *tache noire sclerotique* should not be evident if the death occurred in the water.

Observations

Immediately after recovering the body from water, the eyelids should be carefully opened. If death occurred on land and the body was exposed to air

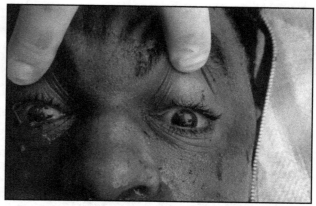

The corneas of this drowning victim remain relatively clear even after eight days of being submerged in 34°F water.

for even a brief time, a noticeable line will be observed horizontally on the eye. This line or border between clear and cloudy cornea, and white and discolored sclera, is normal when death occurred on land and the eyes were exposed to air. It is not normal for a drowning victim. When death occurred in the water, the line or border indicating the position of the eyelids is not present. If there is any clouding of the cornea or discoloration of the sclera, it should be evenly distributed over the visible portion of the ocular globe. Because this horizontal corneal clouding is extremely transient, it must be observed and documented by the public safety diver immediately upon removing the body from the water.

Body Cooling Rate (Algor Mortis)

At the moment of death, all muscular action ceases. Since most body heat is generated by muscle activity, the body's ability to generate heat also ceases. When this occurs, the body then becomes a container at a known temperature (98.6°F/37°C), which will begin to cool at a fixed rate until it reaches ambient (surrounding) temperature. This is the theory; in practice, it is not quite that simple.

There are many factors that will determine the rate of body cooling. Indeed, there are many factors that determine the core temperature of the body at the time of death. Because of the number of variables, many medical examiners put little value in using core temperatures to establish the time of death. The following information is provided to explain the variables and why an underwater investigator might steer clear of calculating the time of death based on core body temperatures.

Calculation of Postmortem Interval Using Rate of Cooling

A rule of thumb states that a body cools at a rate of 1.5°F/0.83°C per postmortem hour (in air). The following formula may be used to determine the postmortem interval:

> Postmortem Interval (hrs.) = (98.6°F – rectal temperature) ÷ 1.5 (°F/hr)
>
> *or*
>
> Postmortem Interval (hrs.) = (37°C – rectal temperature) ÷ 0.83 (°C/hr)

For example, a deep rectal or core temperature of 95.6°F (35.6°C) represents a drop of 3°F (1.7°C), and therefore indicates a postmortem interval of approximately two hours (3°F ÷ 1.5°F/hr, or 1.7°C ÷ 0.83°C/hr). A caveat to this rule of thumb is not knowing the starting temperature of the victim (heavy exercise can raise core temperatures to 104°F), variations caused by ambient temperatures, and environmental conditions affecting conduction, convection, radiation, etc.).

With the above factors in mind, other reputable studies have revealed that a clothed body in air will cool at the rate of 2.5°F (1.4°C) for the first six hours (in the above formula, substitute 2.5°F for 1.5°F, or 1.4°C for 0.83°C). Still other research has come to the conclusion that an average clothed body exposed to an ambient (air) temperature of 60°F (15.6°C) will take 24 to 36 hours to cool to ambient temperature. The rate of cooling will itself decrease as the body temperature approaches ambient temperature.

An investigator should record as much environmental data as possible so the information can be considered at a later time. Water temperature should be taken at the location/depth where the body was recovered.

Shapiro (Temperature) Plateau

To complicate this body cooling rate issue even further, many researchers have documented that healthy individuals, whose times of death were known and accurately documented, retained a normal body-core temperature for the first one to five hours after death (i.e., a cooling rate of zero).

This Shapiro plateau is often referred to as the postmortem temperature plateau. With this knowledge, it is conservative to say that using any standard formula to determine the time of death would be a dangerous practice. At the very least, it would tend to be inaccurate in the early postmortem interval.

Body Cooling in Drownings

A body will cool many times faster in water than in air. In fact, body cooling rates can occur up to 25 times faster in water than in air. Insufficient research and a poor database have rendered inaccurate the determination of time of death in bodies recovered from water. It is extremely helpful to have an infrared electronic thermometer that can provide an approximation of the deceased's body temperature. A rectal thermometer is the most accurate and should be used for a precise temperature measurement.

The Algor Mortis Clock (Determining the Time of Death)

Algor mortis (body cooling) can still be utilized as evidence for determining a rough approximation of the time of clinical death for an individual victim using the following method, which allows the investigator to determine the rate of cooling for the specific body and environment in question.

When the body is located, it must remain in the same medium (air or water) as it was at the time of death. In addition to this, the temperature of the air or water must remain fairly constant throughout this process.

In a body located on land, this usually does present a major problem. A cooling rate for a body recovered in deep, cold water (e.g., 55°F/12.8°C) and transported to shallow warm water (e.g., 65°F/18.3°C) cannot be accurately determined using this technique.

The technique is a simple one: The body temperature is taken as soon as possible after it is recovered. The temperature of the body should ideally be taken with a thermocouple device capable of readings well below normal body-core temperature. Most common clinical thermometers cannot be used in this low-reading range. The temperature reading should be taken in several sites, such as deep-rectal or visceral (deep within the abdominal cavity).

Once recorded, this process should be repeated at 15-minute intervals. After a minimum of four readings, a graph can be constructed of temperature vs. time, and the temperature line extrapolated (extended) backward until it reaches 98.6°F (37°C). Although a minimum of four separate readings are suggested, the more readings that are taken, the more accurate the results of the graph will be.

This technique is perhaps the only safe method of determining the time of death using a body cooling rate graph. Even though it is fairly accurate because it establishes a cooling rate for a specific body under specific conditions, a word of caution should be issued: This technique for estimating the time of death assumes that the body's core temperature was a normal 98.6°F (37°C) at the time of death. This may be a false assumption.

Factors Influencing Algor Mortis

No two bodies will cool at exactly the same rate, even under identical environmental conditions. The following will serve as a guideline. The main purpose of this section is to show the incredible number of variables influencing algor mortis.

Factors That Increase Body Cooling Rate

Water: Water will conduct heat away from the body many times faster than air. Some references quote as much as 2 to 25 times faster.

Currents: Either air or water currents will increase the cooling rate of a body.

Activity: Antemortem immersion in cold water, such as swimming, scuba diving, or even immersion in cold water following an upset-boat incident, may reduce the core temperature. Similarly, exposure to cold air or even being caught in a cool summer rainstorm can effectively reduce core temperature. If death follows a hypothermia incident, the algor mortis graph cannot be accurately extended backward to an assumed 98.6°F (37°C).

Age: Older people with reduced muscle mass, especially in the presence of a reduced fat layer, generate less heat. Many people over the age of 70 (in cooler climates) are chronically borderline hypothermic.

Body Mass Index (Muscle-to-Fat Ratio): The bodies of people who lack an insulating fat layer (skinny people) will cool faster than average. In addition, the mere size of a body will, to some degree, affect the cooling rate. The body of a 60-pound (27.3-kg) child will cool much faster than the body of a 220-pound (100-kg) adult, even though both may have a similar body mass index.

Clothing: A lack of clothing will result in poor insulation to the ambient temperature. To further complicate this, some clothing that has an excellent insulating quality in air loses most of its insulation qualities in water.

Factors That Decrease the Cooling Rate

Body Mass Index (Muscle-to-Fat Ratio): The body of an obese person will cool much more slowly than that of a person with a thin insulating fat layer.

Agonal Struggle: If there has been a remarkable struggle to survive, core temperature may be elevated at the time of death. Even death following a short submergence in frigid water could exhibit an elevated core temperature, especially if death followed cardiac arrest or a fainting spell brought on by the shock of immersion. (Although an elevated body temperature at the time of death does not affect the cooling rate, the body when examined will still be warmer than it would have been had the core temperature been normal at the time of death. Determining the cooling rate in this case will not affect the error resulting from assuming a normal core temperature at the time of death.)

Health: If a viral or bacterial infection is present in the body at the time of death, core temperature values may be elevated. Values in the range of 105°F (40.6°C), in the case of pneumonia or other infections, are not uncommon. Core temperatures in cases of heat stroke may read as high as 110°F (43.3°C). In these cases, postmortem temperature is of little significance for determining the postmortem interval.

Radiant Heat: In very shallow water (less than 10 feet/3 m deep), radiant heat from the sun, especially if the water is clear, can slow the cooling rate. In cases where the victim is clad in dark clothing, this reduction in cooling rate can be quite dramatic.

This may be compounded by the procedure of recovering a body from deeper water and transporting it into direct sunlight on a hot summer day for the purpose of taking a series of core temperatures.

Clothing: Normal clothing has a poor insulating quality when immersed in water. However, specialized clothing such as sailing/exposure suits and diving suits (wet or dry varieties) must be considered differently. These suits have good insulating qualities. Even when completely flooded with water (which is rare), these suits, which are made from an insulating material (closed-cell neoprene, etc.), retain a large degree of insulating qualities. Waterproof clothing, such as rain gear, chest waders, etc., can also be good insulators, since they effectively prohibit water circulation next to the skin.

The Algor Mortis Clock — Is It Accurate?
At best, this algor mortis clock can be used as a rough indicator to determine the time of death for the first 24 hours after death. It is indeed unfortunate that many calculations and estimations as to the exact time of death have been accepted by criminal and civil courts using this clock. Like any tool, the algor mortis clock is most widely used in concert with other observations.

RIGOR MORTIS
Perhaps one of the most frequently observed and least understood (by the public safety diver) phenomena is rigor mortis. Rigor mortis is also referred to as postmortem rigidity. Rigor mortis appears as a gross stiffening of the limbs of a deceased person and often presents a rather macabre sight. Despite this, once its formation and cause are learned, it can be understood that postmortem rigidity is a phenomenon that has its own story to tell.

Onset of Rigor Mortis
In most cases, death is immediately followed by primary muscle flaccidity. All muscles, voluntary and involuntary, in the absence of any nerve impulses or input from the brain or spinal cord, become completely relaxed and soft to the touch.

Rigor mortis, which is a stiffening or hardening of the muscles, usually begins at death, becoming noticeable within two to four hours and continuing until it is fully established in all muscles of the body. Under normal conditions, rigor mortis will be fully established in approximately 12 hours. Similarly, if the conditions under which the body is located are maintained, rigidity will (once fully established) begin to disappear at the same rate at which it appeared.

While this model presents the "simple truth," it does not encompass the whole picture. Rigor mortis is often said to begin at the top of the body (neck) and progress down through the victim, establishing itself lastly in the legs. Likewise, it is said to abate in a similar fashion.

The truth is much more simple. Rigor mortis is first noticeable in the smallest muscles of the body and takes the longest time to visibly manifest itself in the largest muscles (legs). The smaller muscles of the neck are easily tested for rigidity, and in these muscles a degree of stiffness may be felt (by carefully rotating the victim's head laterally) before any stiffness is felt in the arms or legs. By the time the large muscles of the legs and arms have manifested a profound degree of rigidity, the neck muscles may be seen to be quite flaccid. This is due to the fact that postmortem rigidity is in the process of leaving the smallest muscles of the body while it is still progressing in the larger muscles. Simply put, rigor mortis begins and ends in small muscles first and in larger muscles last.

Cutis Anserina (Gooseflesh)
Cutis anserina (gooseflesh or duck bumps) are often noticed on the skin of victims who have been recovered within one to three hours after drowning. This gooseflesh-like appearance is usually noticed on the arms and chest of the victim. Cutis anserina is the first visible manifestation of rigor mortis. Lying directly beneath the skin of the arms, legs, chest, and elsewhere are very small muscles called erector muscles. These muscles once served a very useful purpose: When the skin is chilled, these muscles cause each individual hair to stand on end. This reaction is an attempt by the body to increase the thickness of the insulating layer of hair and thus retain body heat and is a holdover from times when humans had enough body hair to make this an effective response to cold.

Since these muscles are very small, they offer the first site for a visible manifestation of rigor mortis. This gooseflesh is due merely to the rigidity of these tiny muscles beneath the skin. The only diagnostic value of cutis anserina is that it may be considered the first manifestation of rigor mortis.

Rigor Mortis — Its Origin/Cause

Rigor mortis is caused by chemical changes in the protoplasm of muscle cells. During normal cell respiration, carbon dioxide is quickly removed. Laboratory studies have revealed that when muscle pH levels reach 6.6 to 6.3, myosinogen (a protein in muscle tissue) coagulates. Further acidification and time lead to the coagulant being dissolved. This is the stage where rigor mortis then leaves the body. Once rigor mortis has ceased, the body is said to be in a state of secondary flaccidity, and rigor mortis cannot reestablish itself.

Home Experiment — Demonstrating Rigor Mortis

If an ordinary raw egg is broken into a shallow dish, it can be seen to have a design or construction similar to that of a single human muscle cell. The egg yolk represents the nucleus, and the egg white, the liquid protoplasm.

Remove the yolk carefully, leaving as much egg white (protoplasm) behind as possible. This liquid is similar to the muscle protoplasm we have been discussing. To acidify this protoplasm, add an equal volume of clear vinegar and stir slowly to mix. Since the protoplasm has now been subjected to an acid environment (the acetic acid in the vinegar), small particles of a gel-like coagulant will begin to appear within a few minutes. This coagulant will continue to grow and solidify for up to an hour. This simple demonstration is a visible representation of the formation of rigor mortis in muscle tissue and allows a clear illustration of what is happening within the human body after death.

The Rigor Mortis Clock (Estimating Time of Death)

Rigor usually begins to develop within two hours after death and is usually fully established in six to 12 hours. Within 24 to 36 hours, rigor usually begins to leave the body at the same rate it became established.

Unfortunately, there are too many usuallys in this statement to make an effective clock from the rigor mortis phenomenon for use in accurately determining the time of death. Indeed, there are many well-documented cases where widespread rigor mortis was still present in the body 45 to 60 hours after death. Even so, this does not preclude rigor mortis from being a valuable observation for use in determining the time of death.

Factors That Speed the Onset of Rigor Mortis

It is very important to remember that any factors that speed the onset of post mortem rigidity also hasten its rate of disappearance.

Heat: Warmth or heat hastens the onset of rigor mortis. This warmth may be due to ambient conditions such as hot weather, warm water or last known activity (e.g., heavy exercise, agonal struggle, etc.). Death that may have been caused (or contributed to) by heat stroke would also increase the speed at which rigor mortis is established within the body.

Disease: Even simple infections that may result in a fever can sufficiently raise the core temperature of a body to significantly increase the rate of development of rigor mortis.

Agonal Struggle: Rigor mortis is also hastened by a violent predeath (antemortem) struggle. As a result of this muscular activity, lactic acid concentrations may already be high prior to death.

Epilepsy: When antemortem convulsions due to an epileptic seizure have occurred, the onset of rigor mortis may be greatly accelerated. As with agonal struggle, this is a result of antemortem muscular activity.

Poisoning: Many poisons, such as strychnine, cause violent seizures prior to death. These seizures (as in epilepsy) will hasten the onset of rigor mortis.

Muscle Mass: Since rigor mortis is a phenomenon that involves the muscles, a young, well-developed and healthy musculature will manifest a greater degree of rigidity than that of an older person who may possess a much smaller muscle mass.

Factors That Slow the Onset of Rigor Mortis

Temperature: Since rigor mortis is primarily a chemical reaction, a lack of heat effectively slows its onset and departure. Simply stated, a drowning in cold water could account for a much slower onset and departure of postmortem rigidity than one in warm water.

Exceptions

Atypical manifestations of rigor mortis are sometimes seen in accidental deaths. For example, where an individual was hanging on to a float for survival, rigor may be quite well-established in the arms and neck (from straining to keep the nose and mouth out of the water) and not yet evident in the legs.

Examining for Rigor Mortis

The public safety diver, if he is to take the responsibility as a first-line observer/investigator, should examine the body for the degree and development of rigor mortis. Fortunately, this can be done in the privacy of the underwater environment, since once the body is located (in all but the most hazardous conditions), its examination should take place prior to its removal.

Setting the Postmortem Clock

Perhaps even more important than merely observing rigor mortis is the determination of (a) its degree of establishment and (b) whether it is progressing through or leaving the body.

Examination Techniques

To examine the body, the limbs and joints must be manipulated. When moving the body to test for rigor mortis, a basic knowledge of the human musculature is necessary. The following major muscle groups will be discussed in order of their size, from smallest to largest:

Jaw: Careful manipulation of the jaw affects the temporomandibular muscles. These muscles are located at the base (or hinge) of the jaw and extend upward along the side of the face between the eyes and ears, progressing upward to the temples. This is a complex group of many small muscles and represents one of the first sites of rigor mortis' arrival (or departure).

Neck: The muscles of the neck are the next largest muscle group that may easily be tested for rigor mortis. If the victim's head is carefully rotated laterally (side to side), any rigor development in these muscles becomes very apparent. The degree of rotation should be no more than 45° to either side. Immediately following death, the neck becomes extremely pliable, but this disappears as rigor develops in these muscles.

Fingers: The fingers and wrists are controlled by (connected to) the muscles of the forearm. Flexing the wrists and fingers tests for rigor in this muscle group.

Forearm: The next largest muscle group is located in the arms, between the elbow and the shoulder: the biceps and triceps. They may be tested for rigor simply by bending the arm at the elbow.

Feet: The feet, toes, and ankles are controlled by muscles below the knee. Flexing the toes and the feet (at the ankles) tests for rigor chiefly in the lower leg (calf and shin).

Knee: The knees, when flexed or manipulated, will reveal flaccidity or rigidity of the largest single voluntary muscle group of the body: the muscles of the upper legs. Flexing the knees will reveal the degree of rigor in these muscles.

Caution: When testing for rigor mortis, the resulting report should always include the time the test was done and the location and extent of postmortem rigidity observed.

Breaking Rigor Mortis

Often it will be necessary to reshape the position of the body for placement in a body bag, rescue basket, stretcher, or vehicle. When this is desirable, rigor must be broken, i.e., the limb must be forcibly manipulated into a more convenient position.

If rigor mortis is fully established, this is not an easy task, since considerable force may be required to reposition the limb. If rigor mortis is in the process of leaving the body, and a rigor-stiffened limb is moved such that the rigor is broken, it will not reestablish itself in that limb.

Although the term breaking rigor is used, in truth nothing is broken. The limb will resist considerably any efforts to move it if rigor is well-established. As the limb is moved, a rather unique sensation will be felt before resistance is lessened. If care is used, the likelihood of damaging the body by a mere repositioning of a limb is almost nonexistent.

Importance of Observing and Recording Rigor Mortis

Postmortem rigidity, although not a definite or conclusive piece of evidence, can, along with other facts and observations, give important clues as to the cause and time of death.

Posture as Evidence of Cause of Death

Since postmortem rigidity begins to establish itself quickly after death, it can be a very useful tool. Most public safety divers are accustomed to the usual, relaxed (often prone), semifetal position of most drowning victims. This position is assumed because of the buoyant properties of water, the natural forces exerted by the skeletal muscles even when relaxed, and the buoyancy of the lungs (which lie nearer the back than the front). In this position, the arms and legs are usually slightly bent at the elbows and knees. The head is often tilted slightly forward and the spine is slightly curved. When rigor mortis develops, this position is maintained.

Any person who has died on land and remained in a terrestrial environment during the onset of rigor mortis will display a different posture. If lying down, his arms and legs will likely appear rather straight, and the neck will not be tilted forward unless it was supported by an object such as a pillow, etc. The head of an individual who had died on land will also likely be rotated (if only slightly) to one side — a position almost never found in a drowning victim.

As rigor mortis sets in, the body (if it is not disturbed) will become rigid in the position it assumed immediately after death. Hence, a body found underwater that appears to be rigored in a position similar to sitting in a chair would be highly suspicious.

Of course, not all victims are recovered in a typical relaxed position. Frequently when a body is recovered after being trapped in a strong current, its position may be distorted — even bizarre. This rigor-stiffened position, however, should be indicative of the position the body naturally assumed, even if twisted by the current after death. This fact emphasizes the value of photographs and very careful note-taking by the public safety diver before the body is moved. Simply put, the position the rigor has established within the body should be consistent with the position in which the body was found.

Establishing the Time of Death

Due to the large number of variables affecting the speed of the formation of rigor mortis, postmortem rigidity does not offer an accurate clock that can be used to accurately measure the postmortem interval.

A careful observation of the extent and development of rigor mortis, however, should be made and recorded in all instances. When a body is first located underwater, the public safety diver has (in most cases) the time and privacy necessary to make excellent observations. In the privacy afforded by depth and water, the jaw may be flexed along the head (neck), and the arms and legs may be tested to determine the extent of rigor mortis.

When recording the degree of rigor mortis observed, there are many options or scales available. Words such as "light," "moderate," and "full" are subjective and may mean little or nothing to a pathologist. The recommended technique is to rate all muscle groups on a scale of one to ten, with one being no rigor present and ten being (in the observer's opinion) a fully rigored muscle. Using this scale allows an easy comparison between various muscle groups.

Is Rigor Mortis Leaving the Body?

A relaxed jaw and neck, but still immobile arms and legs, would tend to indicate this to be the case. In addition, if a rigor-stiffened limb is manipulated and it does not reacquire any future stiffness, it would be a safe assumption that rigor mortis is leaving that muscle group.

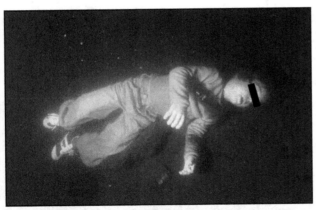

Most drowning victims are found in a characteristic semifetal position. When rigor mortis is established in this position the indication is that the victim entered the water before rigor mortis began to become established. Looking at the position of the body, it is easy to understand how travel abrasions are inflicted on the face, knuckles, knees, and toes of a drowning victim. The adult is being held with his back against the current. A length of steel cable emerging from the river bottom stopped further travel of the body in the 3-mile-per-hour current. The child shown here rests on the bottom in still water.

Is Rigor Mortis Progressing Within the Body?

Should the jaw or neck provide some resistance to movement and the arms and legs show no (or very little) resistance to flexing, then it would be safe to assume that rigor mortis is actively progressing. Similarly, if a rigor-stiffened limb is moved and it continues to develop postmortem rigidity, then rigor mortis is likely in the formative stage.

Caution: In drownings involving cold water, a certain degree of muscle rigidity is often present not only immediately after death, but before death. Dr. Martin Nemiroff, who has been hailed as the discoverer of the cold-water near-drowning phenomenon, comments that in many individuals who have been successfully resuscitated, muscular rigidity (one of the classic signs of death) was present when the rescue was made.

To confuse the matter further, it should be noted that rigor mortis is often absent or nearly absent in the bodies of very old victims or very young babies.

A Fable Put to Rest

This section on rigor mortis would not be complete without putting an old superstition to rest: Stories continue to abound about people "sitting up in their coffins" and drowning victims "sitting up on the autopsy table." These are not true; although this seems to be a self-perpetuating story, its origins are lost in obscurity. It has simply never been documented.

Rigor mortis refers to the hardening or stiffening of muscle fibers, not to their contraction. While there may be a degree of flexion (bending) of the fingers after death, this appears to be the result of physical contraction of the tendons that connect the bones of the hands with the large muscles of the forearm. This contraction is suspected of being a side effect of cooling. This may explain why the hands of many drowning victims are often found to be closed on recovery of the body.

Cadaveric Spasm

Cadaveric spasm is often referred to as instantaneous rigor or cataleptic rigidity. When it manifests itself in the hands, it is sometimes referred to as the death grip.

Cadaveric spasm is exactly what the name implies — a spasm found in the cadaver (body). It resembles rigor mortis because it is a firm and pronounced stiffness or firmness in a muscle group. It is different from rigor mortis, however, since it does not progress evenly through the body at a steady rate.

Cadaveric spasm occurs instantly at the time of death and is confined to groups of voluntary muscles only. Instantaneous rigor is usually noted in the hands, especially when a drowning victim's last conscious effort was an agonal grip on a lifesaving device or a valued possession.

Cadaveric spasm has been reported in many forms. A scuba diver, after passing through the propeller of a large boat and being killed instantly, still held the rubber mouth-grip of his regulator firmly in a set of rigor-locked jaws. Interestingly enough, the forces exerted on the body actually pulled the regulator from the mouthpiece while the victim's grip held the latter securely in his mouth.

A rather common occurrence of instantaneous rigor is often seen when a drowning victim is recovered still clutching his eyeglasses or any personally valuable object in his hand. Instantaneous rigor may also be observed when a body has been recovered from a drowning and is found to be clutching bottom debris or vegetation.

Cadaveric Spasm — What Does It Indicate?
Since cadaveric spasm occurs virtually instantly and only in groups of voluntary muscles, it is only an indicator of what the victim was thinking or doing at the instant of death.

Expensive prescription eyeglasses clutched in a drowning victim's hand may merely represent the emotional attachment (importance) placed on the eyeglasses by the victim. Like the scuba diver who was drawn into the ship's propeller and still (after death) retained his mouthpiece, cadaveric spasm reveals only the actions of the deceased during his last seconds of conscious life.

Cadaveric Spasm — Its Cause/Origin
The physical or chemical mechanism behind cadaveric spasm is not entirely clear. It is possible that a lactic acid buildup in specific muscle groups could hasten rigidity, but not to a degree that would cause it to develop instantly. There remains no explanation that clearly accounts for this phenomenon. Indeed, in some circles cadaveric spasm is still believed to be a myth perpetuated by the novelist.

Certainly, cadaveric spasm is not a myth, and the public safety diver who remains active in this profession will eventually witness it.

Cadaveric Spasm — Its Importance
Cadaveric spasm is formed only under conditions of extreme mental stress and indicates the victim's last thoughts and actions.

When bottom debris or vegetation is found tightly clutched in a suspected drowning victim's hand(s), it is proof that he was alive when he reached the bottom.

Caution: As rigor mortis becomes established, the fingers tend to close slightly and bottom debris may be seen to be clutched in the hand. Cadaveric spasm is quite dramatic. The hand will be frozen in a clutching position even if rigor mortis has not yet developed elsewhere. Finally, cadaveric spasm cannot be simulated after death.

On the Scene: Cadaveric Spasm Indicates Accidental Death
A fisherman was sitting on the stern of a large pleasure yacht, trolling. He went missing during a period of approximately five minutes while his two friends were inside the boat's cabin pouring coffee. The last-seen point encompassed nearly one-quarter mile (400 m) of shoreline.

Upon interviewing the friends, it was learned that the missing man was the only person fishing (trolling) at the time he went missing. It was further learned that he insisted on sitting in the stern of the boat against all advice. In addition to this information, it was also noted that the fishing rod he was using had been a birthday gift from his wife the week prior to the accident. This was the first time he had used it, and the way he had boasted about it tended to indicate quite clearly that he felt it to be of great value.

In this area, it was customary to troll with approximately 150 feet (45 m) of line out behind the boat, and during the time when he had gone missing, the boat had passed within 50 feet (15 m) of a point of land. The owner of the boat recalled the alarm on his boat's depth sounder, indicating a reef. When charts were consulted, it was revealed that the depth of the water ranged dramatically from 200 feet (60 m) deep to a series of shoals that rose vertically to within 15 feet (4.6 m) of the surface in this area. Underwater visibility was less than 2 feet (60 cm).

The public safety divers realized the significance and importance of the fishing rod and began their search of a quarter-mile (400 m) of shoreline. The search consisted of pairs of divers swimming underwater, perpendicular from shore out to a distance of approximately 100 feet (30 m). Holding their arms ahead of them in a plowlike manner, it was hoped that they would intersect the monofilament fishing line. The area to be searched would have presented a nearly insurmountable problem for conventional search patterns. It was hoped that once the fishing line was found, the rod could be located easily and the missing fisherman would be nearby.

On the second pass a monofilament fishing line was encountered by two divers. Approximately 20 feet (6 m) down the line they encountered the lure, firmly snagged in a small rock crevice. A 120-foot (36-m) swim in the opposite direction (along the line) revealed the fisherman — still holding the rod his wife had given him. A deep wound on his left temple was believed to have been caused by the propeller of the boat.

In this case, instantaneous rigor not only assisted the public safety divers in finding the body but also helped rule out any suspicion of a fight or altercation between the victim and the survivors who remained on the boat.

LIVOR MORTIS

Livor mortis is the term used to describe the color change of a body after death. This color change is attributed to the cessation of circulation and a pooling of the blood within the body. Livor mortis is also referred to as postmortem lividity, hypostasis, and postmortem suggillations.

Livor Mortis — Its Origin/Cause

Immediately after death all cardiovascular activity (circulation of the blood) ceases. When this occurs, blood that normally circulates throughout the body gravitates to the lower areas. When this occurs, capillaries in these lower areas become swollen with blood. A small number may even rupture, forming tiny spot-like hemorrhages.

Similarly, areas of the skin that are compressed by gravity or by tight clothing such as a belt, watch band, bra, etc., will exhibit a compression phenomenon wherein the capillaries will have virtually all blood squeezed out of them. This area of compression will appear very bleached or white. When the

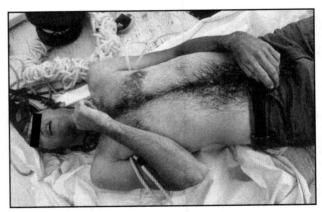

This victim of a fatal swimming accident was recovered by public safety divers 85 minutes after drowning. No rigor mortis was noticed. However, his right arm (elbow, wrist, and fingers) exhibited instantaneous rigor or cadaveric spasm.

compression of an area is due to the weight of a body on a firm/hard surface, it is referred to as contact lividity.

Livor Mortis — Its Appearance

The most outstanding feature of livor mortis is the dark purplish (livid) coloring that accompanies it. In deaths involving carbon monoxide, the predominant color of lividity is cherry red. If cyanide poisoning was the primary cause of death, livor is often seen to be a bright scarlet red or a light pink. This pink coloring is also often common in bodies recovered from icy waters or snow.

The simplest way to explain the appearance of postmortem lividity would be to consider a death on land, when the victim was sitting upright in a chair. The head (face and ears) would appear very pale, especially if the head were tilted back against the back of the chair. The forearms, resting on the lap, would appear a darker purplish color, and the lower legs would be very dark, purplish, and perhaps swollen. The feet and ankles would exhibit the greatest degree of lividity and swelling from the accumulated pressure of the blood settling within the body.

Livor mortis occurs within 30 minutes to 2 hours after death. It reaches peak intensity at 8-12 hours and can shift position on the body depending on movement and position. It becomes fixed or unmovable by 10-12 hours. It is important to know that higher ambient temperature will accelerate the fixation, whereas cooler temps (like those found in

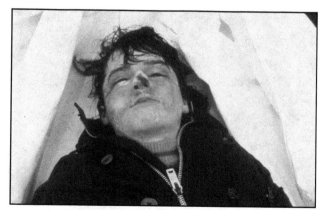

Contact lividity is not commonly seen on a victim recovered from the water. When it is noticed, it is generally seen around the face. This youth was found underwater, lying face down on a rocky bottom. The weight of his head on a rock produced contact lividity in the area of his eyes and nose.

cold water) will delay fixation. For this reason it is important to record the ambient temperature (water and air) at the time of recovery.

TARDIEU'S SPOTS

In areas where the collection/congestion of the settling blood actually ruptures the tiny surface capillaries, continued pressure in this area may lead to a discoloration often referred to as Tardieu's spots. These spots are merely tiny spot hemorrhages and appear as visible but small circular skin hemorrhages. Tardieu's spots are almost always seen in the lower areas of the body. Suicidal hangings, where the body often remains in a vertical position for several hours before being discovered, almost always exhibit Tardieu's spots in the area of the ankles and feet.

Although Tardieu's spots are often associated with (and once thought to be proof of) asphyxial deaths, they should not be interpreted as indicating any specific cause of death.

CONTACT LIVIDITY

The term contact lividity is a contradiction in terms. Where a body is in contact with a hard surface, blood is squeezed out of the surface tissues. This area is not livid — it appears white.

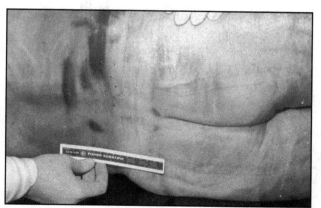

Contact lividity is caused by the weight of the body on a firm surface, compressing the skin and forcing stagnant blood out of the tissues and capillaries. It is often dramatically contrasted with normal postmortem lividity (coloring) of the tissues where blood was allowed to pool. During autopsy, this individual who died on land was turned over. The darker areas represent contact lividity. He was lying on his back after death, allowing the blood to pool in the areas that were not in contact with the ground. Obviously his buttocks supported much of his weight. The two large horizontal bruises just above the buttocks matched exactly to the height of the hood of the suspect hit-and-run vehicle, which was later impounded.

FACTORS AFFECTING THE DEVELOPMENT OF LIVOR MORTIS

If the deceased suffered from chronic anemia and had a resultant low hemoglobin content in his blood, lividity will not be obvious. Similarly, if death was a result of an acute blood loss, lividity may be drastically reduced.

Dark-skinned individuals will also tend to display a reduced degree of lividity. Although lividity will still be present, it may be less visible.

LIVIDITY IN DROWNING VICTIMS

Since the weight of a drowning victim is countered by the buoyant force of water, lividity and contact lividity should be nearly missing. Many sources state that, in drowning victims, lividity is usually pronounced in the areas of the head/face and hands, due to the semi-fetal face-down position so common to drowning victims. In truth, lividity in drowning victims is usually absent. There are two exceptions, however. These are:

1. **Contact lividity.** When the victim is lying in a face-down (prone) position, whitish areas may be seen on the knuckles or back of the hands, as well as on the face. These areas of contact

lividity represent areas where the hands or face were pressed (even if very gently) against a hard object such as rocks, gravel, logs or small sticks. These materials are often found on the bottoms of lakes, rivers and ponds, and will often result in contact lividity being present in drowning victims recovered after two hours of submergence.

2. **Current lividity.** Where a drowning victim is held trapped in a strong current, the pressure exerted by the current may push the stagnant blood to the extremities, which are located downstream. Conversely, the areas of the body that are located on the upstream side may appear pale or white because the pressure of the water against the surface tissue has forced the blood from this area.

THE LIVIDITY CLOCK

Lividity is ordinarily visible within 30 minutes to four hours after death. It is usually well-developed in eight hours and fully established in 12 hours. Knowing this, it then becomes easy to understand that if a body is recovered on land and no lividity is observed, but in an hour or so lividity is visible and progressing at a regular rate, death was likely within the preceding hour.

Unfortunately, as it has already been discussed, postmortem lividity is rarely observed to any great degree in bodies that were alive during submergence, other than contact/compression lividity or current lividity.

In the early postmortem interval, lividity is not fixed; i.e., if a body is moved to a new position, lividity will reestablish itself. Once lividity becomes fixed (in 12 hours under normal conditions), repositioning of the body will not have any effect on the original pattern of lividity.

Lividity alone, as with other postmortem observations, does not give us an accurate clock with which to determine the time of death or the postmortem interval. Fixation of livor has been noted within 30 minutes of death (though rarely), and cases have been recorded where it has been absent 12 hours postmortem.

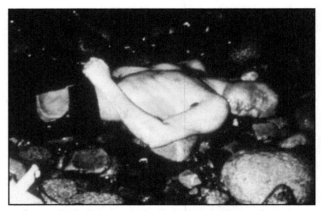

Current lividity is caused when a body is held securely against a fast-flowing current. Instead of gravity causing the blood to pool, the current squeezes the blood in the direction of its flow. The hands are uncharacteristically white, yet the head is livid. The current was obviously flowing in a direction from the feet to the head. This is corroborated by the fact that his shirt has been removed and his pocket turned inside-out (by the current). This youth had been missing for approximately 10 hours.

IMPORTANCE OF LIVOR MORTIS

Even though there are many variables, livor mortis furnishes two important types of information:

1. **Time of death.** The degree and development of livor mortis will help establish the time of death, and although not conclusive itself, it can be used to approximate the time of death and support other observations. Other observations may include rigor mortis, witness statements, etc.

2. **Cause of death.** If a body is recovered underwater exhibiting a well-established degree of livor mortis, then it may be assumed that the victim died perhaps several hours prior to submergence. The distribution of lividity will (if death occurred on land) indicate the resting position of the body postmortem.

LIVIDITY VS. CONTUSIONS

Postmortem lividity should not be confused with antemortem (before death) contusions (bruising). Postmortem lividity is usually less localized in appearance than a bruise, and once one is familiar with the appearance of lividity, it is fairly easy to differentiate it from a blunt-trauma injury.

Postmortem lividity involves blood that has settled in the body due to gravity. This blood is still mostly contained within the capillaries. If clotting has not yet taken place, pressure on the areas exhibiting livor (using a knuckle or finger point) will result in a localized blanching (whitening) of the area.

If the coloring is due to antemortem bruising, the pressure test will have no effect because the red blood cells, which give color to the bruise, are trapped in the tissue spaces and cannot be easily evacuated.

On the Scene: Wounding Prior to Death

A seven-year-old boy, wet and tearful, appeared on foot at a house near a large pond. An incomplete story related by the tearful youth resulted in the local public safety dive team being summoned to search the pond. After a 12-hour search in the murky water, the body of an adult male was recovered. An area of his forehead just above the eye exhibited an obvious discoloration. This discoloration was highly localized. Upon seeing this, the investigators assumed it was a result of livor mortis, but the dive team suspected injury sustained prior to death. Upon autopsy, it was discovered that this was indeed a bruise from an injury that indeed occurred just prior to death.

The child's story began to make sense. Finding himself in trouble in the pond, the boy called for help. From a high embankment (approximately 10 feet (3 m) above the water) the man had thrown an old wooden door in the direction of the boy. After throwing this door into the pond, the man then dove into the water. The door reached the child, but it is unclear whether or not the would-be rescuer ever did.

The child's story, along with the discovery and identification of the bruise (not livor mortis, as was suspected) ended all suspicions of a foolhardy summer swim by the man and child. Indeed, it elevated the status of the deceased to that of hero. He died saving the life of a young boy, and the man was a nonswimmer.

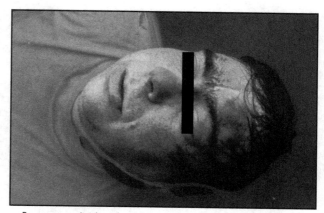

Postmortem lividity should not be confused with injuries. This contusion to the area above the left eye was a result of any injury received at the time of death. Note: There is not observable lividity elsewhere on the face. Victims recovered from the water rarely exhibit lividity to this extent, and when discoloration is noticed to be the severe, an injury is usually suspected.

Physical Wounding of a Body

In many cases, when a body is recovered and inspected closely, it will exhibit what is commonly called wounding. This wounding or damage to the body may help to reveal the cause of death. If for no other reason, the body should be examined very closely by the public safety diver. Valuable clues that are often missed by the casual observer can serve to raise the caution flag.

Every body recovery should be first investigated as a homicide; then, as the facts become clear and point to a cause of death, foul play may be ruled out, leading to a conclusion of death due to accident or suicide.

The importance of good powers of observation, accurate and complete notes, and photographs where possible cannot be overemphasized. The task of finding the body is the first step; observing, deciphering, and recording the find completes the investigation.

There are many ways in which the human body may be wounded or damaged. Was the physical wounding inflicted before death, after death, or did it cause the death? If it was not severe enough to cause the death, could it have contributed to the drowning? Are the wounds self-inflicted or the result of an attack by another person (or animal)? These are but a few of the questions that must be answered. Surely it is the responsibility of the chief investigator, coroner, and forensic pathologist to answer these questions, but good observation and reporting skills by the public safety diver can often lead others to a more detailed investigation of a homicide. Conversely, suspicious physical damage to a body can often be explained as natural, shifting the avenues of investigation away from a possible homicide to a suicide or accidental-death investigation.

There are a variety of ways in which the human body may be damaged before death, after death, or at the moment of death. This portion of the manual is not meant to be a compendium of all possible injuries but rather a simple and frank discussion of what the public safety diver/investigator should look for in most water-related deaths.

There are three time frames when a body may be wounded (injured/damaged). They are antemortem (before death), postmortem (after death), and finally agonal, or injuries that are received at the moment of (or very close to) death.

Antemortem (Before Death) Wounding

Any wounds inflicted on the body before death will be a determinant in ascertaining the cause of death. Simply put, wounds found to be antemortem in nature will help the investigator to ascertain the actual cause of death. This cause of death includes the determination of death due to homicide or suicide.

Defense Wounds

Defense wounds are defined as those wounds that are inflicted upon a living person as he used any part of his body or limbs to defend himself from attack.

Defense wounds are most commonly seen in the areas of the hands and/or forearms, since it is an instinctive reaction to raise one's arms in defense against an attack. Defense wounds most commonly appear on the palms and palmar surface of the fingers. On the forearms, defense wounds would most often appear on the inside of the forearms of either arm. Severe lacerations in these areas should lead the investigator to suspect foul play.

Caution: On occasion, injuries in the palmar region of the hands are noticed that are not indicative of classic defense wounds. Classic deep lacerations

on the palms and forearms from a knife-wielding assailant must be differentiated from wounds received while sliding down a steep embankment into the water. These survival wounds will usually not appear as singular deep lacerations. Instead, they will have a complex striated pattern that indicates more of a severe, deep scuffing injury than a deep cut. In survival wounds such as these, there may also be some wounding to the elbows, knees, and face, especially the chin, as a fall down a rocky incline is accelerated. In addition to this damage, fingernails may appear to be dirty, damaged, or partially missing. In severe cases, the face may appear to have been injured by a blunt trauma, and teeth may have been partially or wholly dislodged.

In drownings where the victim had been holding onto an object to save his life, the hands may be injured, especially when the grip was for survival and/or the object was irregularly shaped or sharp. It is the task of the forensic pathologist or forensic laboratory technician to determine the exact mechanism and cause of all wounds; however, the public safety diver often possesses a great deal of experience in these areas, and his experience combined with his critical observations are often of primary importance.

Agonal (At the Time of Death) Wounding

In most cases, wounds that are sustained at or very close to the time of death may be obvious. Head injuries sustained as a result of a fall will usually be apparent to the public safety diver as he recovers the body, but the investigative diver should be able to look beyond the obvious and discover details that would be expected only of a seasoned veteran.

Determination as to whether an injury was sustained seconds or minutes prior to death, or very shortly after death, is usually the task assigned to the forensic pathologist. This determination is often made only after the site of the injury has been carefully examined, often with the use of a microscope.

The procedure for determining the status (before or after death) of an injury is often a simple examination to ascertain the amount of bleeding present at the site of the injury. This simple test, however, falls short of 100 percent accuracy. While localized swelling was once deemed to be proof of a viable, active circulatory system, it is now known that soft tissues may swell remarkably if the site is located in a position where blood could pool. Perhaps one of the most reasonable procedures to follow in determining the time frame of an injury is to consider how the injury or injuries were sustained.

Falling Injuries

Falling injuries are quite common in accidental and suicidal drownings. These injuries are sometimes found on victims who have either fallen (or dived) into shallow water and struck the bottom. Falling injuries may also be documented on individuals who have struck an object prior to entering the water.

Most falling injuries are a combination incision/blunt trauma. This is caused by the force of the blow due to the momentum of the victim's body as it struck the object. Injuries such as these are usually quite dramatic and may have actually caused the death, independent of submergence in the water. Where there was a massive head wound and a subsequent instantaneous cessation of heartbeat followed by submergence in water, the wound will often show very little signs of bruising or bleeding. This is because the water has cleaned the wound, and in the absence of heartbeat, little or no hemorrhage took place into the surrounding tissues.

Caution: There is no strict rule equating the appearance of a wound with its chronology in the accident. While in many instances a clean wound would seem to indicate the lack of circulation of blood, the site of the wound is also important in determining its status. A wound to the top of the head is easily washed and cleansed after death, especially in the presence of a current, and will often appear as though it was received at the time of death or after death occurred. Conversely, a serious wound inside the mouth may not be as exposed; hence, it may be seen to bleed when the victim is removed from the water. This is due to the simple fact that the closed or partially closed mouth held the blood trapped until the body was recovered. At the very least, it prevented water circulation in the area and inhibited the cleansing effect of the water.

When recovering a body, a careful inspection should always be made for injuries. Although decomposition was in an advanced stage, the crushing injury on the top of the victim's head indicated either a blow or a falling injury. The pathologist later confirmed that death was due to a low-speed impact. Very little skull damage was noted, but the victim's neck had been broken. His body was found in shallow water beneath a cliff.

Damage Due to Boat Propellers

In most cases, damage to a body due to boat propellers will be in one of two major sites, depending on the orientation of the victim's body when struck.

Postmortem Propeller Damage

Many bodies eventually resurface after a period of submergence. This is due to gas production within the body. When this refloat occurs, in most instances the body is floating in a face-down attitude. If the body is damaged at this time by a boat's propeller, damage will nearly always be to the back and shoulders, back of head and neck or buttocks.

Antemortem Propeller Damage

When a swimmer (or a person who accidentally falls from a boat) is hit with a propeller, he is usually in a vertical or swimming attitude, attempting to avoid collision with the approaching boat.

It is an instinctive, last-minute reaction for a swimmer to raise his arms to ward off an approaching boat. Propeller damage in these cases will involve the arms, in particular the forearms and hands. Since a person's arms have little effect on a rapidly approaching boat, and resistance from the water will prevent a person from being thrown clear, propeller damage is usually also found on the face or side of the head.

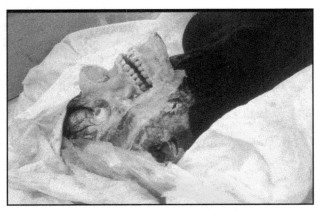

Only the public safety divers had noticed the injuries on this victim's cheek. It was obvious that he had been wounded by a propeller while alive and conscious. A similar wound to his left hand (palmar region) confirmed this wound was antemortem (before death). It is the duty of the public safety dive team to carefully inspect, observe, and record their findings at the termination of any body recovery.

Propeller marks usually appear as a series of parallel striations or deep cuts on the victim. Smaller, high-speed propellers tend to leave cuts that are very close together, while propellers of larger craft (tugboats, etc.) tend to space the lacerations further apart, with each incision often being several inches/centimeters deep.

Other Injuries

The public safety diver should always be aware of any injuries to the victim, no matter how slight. It is important to realize that any recent injury likely played a role in the drowning. Even drownings that are entirely accidental, if investigated in detail, will tend to make sense and yield answers to questions that might otherwise have haunted law enforcement agencies and the next of kin forever.

On the Scene: Cut Finger Explains Drowning

A six-year-old boy had fallen from a wharf. The drowning was not witnessed, and the local dive team was called only after the child had been missing for several hours.

Upon locating the body of the youth, it was noticed that a deep cut completely encircled his left index finger. His fishing rod was found on the bottom within reach of his body.

It is believed that the youth had encircled his own finger with the fishing line in an attempt to free the hook from a snag. The hook and bait were still snagged on the dock piling when the fishing pole was recovered. Probably in an attempt to pull the hook free, the youth cut his finger deeply. The resulting pain and confusion likely caused him to lose his balance and fall into the water. He had previously been warned by his father not to stand so close to the water. The child was not wearing a personal flotation device.

On the Scene: Cut Foot Explains Drowning

A nine-year-old girl was last seen walking barefoot on the wharf at a small marina. She was water wise but not wearing a personal flotation device. The girl could not swim.

After realizing his daughter was missing, the father conducted a search of the area but could not locate her. He called the local police, who in turn contacted their public safety dive team. Three hours had passed.

When the body of the girl was brought to the surface, a deep cut was noticed on the outside of her right foot. Her body had been located immediately beneath the right edge of the wharf where she had last been seen. This deep laceration was measured to be approximately 2 inches (5 cm) from the sole of her foot, at the base of her ankle. A physical inspection of the wharf at the site where her body was found revealed a large spike protruding from the rail, 2 inches (5 cm) from the deck of the wharf.

Excellent observation and documentation by this dive team made reconstruction of this mysterious accident a simple matter. In this case, the dive team had chosen not only to recover a body but also to investigate the accident.

On the Scene: Investigation Precedes Search

A 57-year-old man who was known to be living on his boat in a large marina was not seen by the marina security guard during his morning rounds. This was unusual, since they were friends and the guard was in the habit of visiting him for early-morning coffee.

The man was a heavy drinker and had no known next of kin or friends other than those who knew him at the marina. The guard who had worked the previous night was contacted, and he stated that he had seen the individual at approximately midnight, when he had staggered up to the garbage bin to dump his empties. The garbage bin was located at the top of the wharf ramp, a distance of approximately 300 feet (90 m) from his boat. When this man was last seen, it was snowing lightly, and about 2 inches (5 cm) of fresh snow had fallen. He was quite intoxicated.

This search presented some unique problems: cold weather, a large search area, and profuse bottom debris. In addition, it was not even known if the individual had indeed fallen into the water. Upon further questioning, the dive team learned from the night guard that the snowfall had ceased shortly after he last saw the missing person.

The first public safety diver walked the route to the individual's boat slowly. He noticed that approximately 2-3 inches (5-8 cm) of snow had fallen during the night and that this snow had created a uniform ridge over the tie bar. This tie bar was a small ledge/railing that was mounted approximately 4 inches (10 cm) above the edge of the wharf. It was fitted with large metal eye bolts that were used to secure the boats.

A short distance along the wharf, on the left side, this neat uniform snow ridge was markedly lower for a distance of approximately 3 feet (1 m). This coincided with another line that a thoughtless boater had tied to the far rail; i.e., this line stretched from the bow of a boat across the wharf to an eye bolt on the far side and had to be stepped over in order to pass.

A careful dusting away of the snow from the rail revealed a few small blood splatters. Inches/centimeters from the blood splatter was a spike that protruded approximately 2 inches (5 cm) from the rail. Photographs were taken at the site of this fresh red (still frozen) blood, and the search was commenced at that point.

The victim's body was located directly beneath that location. Careful inspection of the body revealed a deep gash in the area of the right temple. The subsequent autopsy revealed that the middle temporal artery had been severed. The cause of his death was officially recorded as "accidental drowning involving alcohol." His blood alcohol level was 0.26 percent. According to the pathologist, the blow to his head, combined with his

After noticing a slight depression in the snow, it was carefully dusted away to reveal a small blood splatter. The spike (eye bolt) to the right of the blood splatter is believed to have caused the injury. The victim's body was found in 25 feet of water directly below the blood splatter.

After one hour dive planning and preparation, the victim's body was recovered in less than two minutes bottom time. Note the foam that is being removed from the water. The red-pink appearance of the victim is often observed in individuals who drown in cold water. The water temperature in this marina was 38°F.

intoxicated state and the shock of the cold water (38°F/3.3°C), coupled with the fact that he was a poor swimmer, likely caused him to slide below the surface of the water without even calling for help.

The public safety dive team engaged for this search chose to act slowly and deliberately to investigate the accident in preparation for their underwater search. They had truly evolved into an investigative dive team. (The boat with the overextended mooring line was moved the following day.)

Postmortem (After Death) Wounding

It is very important for the public safety diver to understand and to be able to identify various forms of postmortem damage or wounding that may be observed on any body recovered from the water.

The wounding of or damage to a body after death may conceal trauma or injury that was inflicted prior to death. If the postmortem damage is severe and the investigator not astute enough, then it could very well happen that foul play would not be discovered until examination on the autopsy table. Of course, any delay in a death investigation results in other evidence being lost and the likelihood of a successful outcome reduced. Conversely, certain aspects of postmortem wounding, to the uneducated eye, may look suspicious and lead to countless hours spent investigating a false lead.

The investigative dive team should be able to look beyond the sometimes horrific appearance often caused by nature. Being able to identify what is natural and what is unnatural is the main task of the investigative diver when considering postmortem wounding of a body.

Because of the many sources and types of wounding a body may be exposed to, this area of investigation may be best examined by considering the various sources of wounds or damage that could reasonably be expected.

Travel Abrasions

The normal position that a body assumes after drowning or being submerged in water is face-down (prone) with the spine slightly bent forward. In addition to this, the legs may be slightly drawn in and the arms slightly flexed.

This semifetal position allows the following areas to be in contact with the bottom of the lake, river, or ocean.

   a. Forehead and nose

   b. Knuckles of either hand

   c. Elbows

   d. Knees

   e. Tops of feet and toes

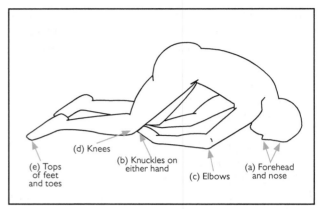

Drowning victims are often found in a face-down, semifetal position. Movement while in this position would likely result in travel abrasions to the forehead, nose, knuckles, knees and/ or toes.

When a body is moved along the bottom of a river, ocean, etc., by currents or wave action, these areas may be subject to erosion. This erosion or wearing away of the skin creates what are commonly referred to as travel abrasions. On occasion, a body may exhibit travel abrasions even when it has not traveled a great distance. The mere rocking back and forth on sand or a gravel bottom (in the presence of wave action) may result in travel abrasions being inflicted in a time period of less than one hour.

Travel abrasions should not be confused with defense wounds, which are usually observed in the palmar (inside) region of the hands and the inside of the forearms. Travel abrasions usually appear as scuff marks on small areas of skin that have the appearance of simply being worn away. Defense wounds usually appear as deep cuts, punctures, or blunt trauma injuries.

In addition to these observations, travel abrasions, upon close examination, will often contain small particles of sand or grit — the medium that inflicted the wounds.

Postmortem Skeletal Injuries

Fractures, or even on occasion complete breaks, of long bones after death may occur in elderly or debilitated persons who have been confined to bed for a long period of time. In cases such as this, an advanced condition known as osteoporosis may cause an increased fragility of bones.

Injuries sustained while traveling down a river after drowning may cause complete and clear breaks in any part of the skeletal system.

In addition to these factors, rough handling by those responsible for the body recovery, as well as the handling and transportation to the hospital/morgue, could easily result in fractures to the spine or ribs.

In cases involving advanced osteoporosis, any attempt to straighten a rigor-stiffened limb could result in a fracture of the arm or leg.

Of particular interest to the investigator is the likelihood of fractures of the ribs in victims where aggressive cardiopulmonary resuscitation has been attempted. In these cases, it is not uncommon to see these fractures accompanied by lacerations of the upper abdominal organs, including the liver. There have been many recorded instances where aggressive cardiopulmonary resuscitation has even led to considerable hemorrhage into the pleural (lung) and abdominal cavities. This could be quite misleading to a pathologist unless he were informed that lifesaving measures such as CPR were attempted.

This fact should be kept in mind, particularly when the victim has been recovered from a motor vehicle accident after a violent entry into the water. A trained and experienced pathologist should be able to differentiate between these (CPR-attempt) injuries and those caused by forceful impact with a steering wheel. Informing the pathologist, however, of the history of the accident and subsequent actions will assist greatly in a correct diagnosis.

Anthropophagy

The term anthropophagy refers to the ingesting (eating) of a human body by any multicelled organism. In the strictest of definitions, this could include large animals, fish, reptiles, crustaceans, etc. Anthropophagy also includes cannibalism. The related word anthropophage (plural anthropophagi) literally means man-eater.

Importance of Observing Anthropophagy

The possibility of encountering incidents of anthropophagy in nearly all bodies of water and areas of the world is nearly certain for the active public

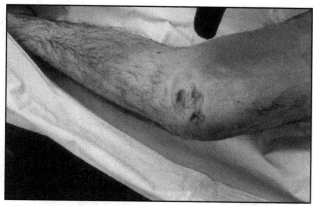

Travel abrasions appear on the knee of a victim whose body was moved only a short distance by water currents. He had corresponding injuries on his other knee, knuckles, and face.

Travel abrasions on this victim were enlarged by saltwater shrimp. Anthropophagy often begins at the site of an injury.

safety diver. Anthropophagy is often shocking to the novice public safety diver, and as such it tends to alter or even render useless a critical, objective investigative mind. Indeed, it is understandably difficult to expect anyone to remain unemotional and detached when recovering the body of an individual who has been half-eaten by nearby creatures.

The public safety diver should remain objective, however, even curious with regard to postmortem artifacts. Anthropophagy in reality is not horrific, obscene, or unnatural. On the contrary, it is natural when considering the ways of nature. It should not be looked upon with an attitude of disgust or scorn, even though its mere presence may give rise to deep-seated, instinctive feelings.

Anthropophagy in all its macabre detail is merely a postmortem artifact that should be carefully observed, studied, and reported on in a clear objective manner by all public safety investigative dive teams.

When any animal dies, its physical return to the elements must take one of two routes: putrefaction or ingestion by other animals (scavenging, or if the animal has not died of other causes, predation). Putrefaction is the consumption (ingestion) of the carcass by bacteria. When this route is followed, many bacteria that will happily feed on a living body may flourish. That is, putrefaction may be hazardous for other healthy living organisms. Putrefying bodies carry disease. Scavenging is nature's way of intervening and safely cleaning up the environment before dangerous bacteria find a breeding ground.

Any life form that depends on animal protein for survival may be anthropophagic. Animals who kill and eat their prey are referred to as predators. Those who eat what has died of other causes are referred to as scavengers; the term carrion-eater is also commonly used as well as some other unflattering terms. Animals who are scavengers clean up before bacteria find a place to multiply. Scavenging is a natural phenomenon; without it, disease would spread rapidly throughout nature. People tend to take a dim view of it, however, when the carrion being scavenged is a human body.

While it has been suggested that the public safety diver should remain objective and curious about a phenomenon such as anthropophagy, this is not easily done. Anthropophagy often takes on a horrific appearance. Anthropophagy can not only alter our perception and objectivity in an investigation, but it also has the ability to physically alter any disfigurement, wounds, or trauma in the victim. For this reason, it should be observed and reported on in an accurate, clear, and concise manner.

Crustaceans

Perhaps among all creatures, crustaceans are to be considered the most voracious. These greedy, active feeders are most adept at producing the phenomenon of anthropophagy, since they depend on their skills as scavengers for their very survival.

Crustaceans responsible for anthropophagy are chiefly, though not exclusively, crayfish, lobsters,

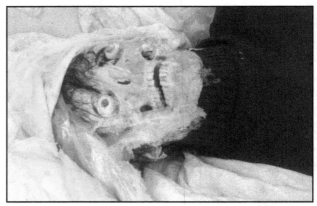

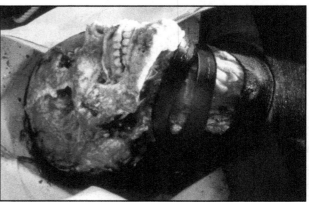

Anthropophagy usually begins at the softest tissue sites or areas of tissue damage. Both bodies were recovered less than 16 hours after drowning. Anthropophagy was due to saltwater shrimp. The bodies were recovered in less than 20 feet of water.

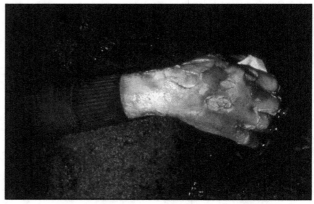

Anthropophagy as a result of fish feeding was noticed on this victim's hand. The body was recovered from a mountain river, and small immature salmon were seen actively feeding on areas of flesh that had loosened.

crabs, shrimps, prawns, and the family of invertebrates commonly referred to as sea lice. They are most abundant and active in shallow, warm waters.

Crustaceans are well equipped to search for and find any body that settles in their territory. They are equally well equipped to tear off small portions of flesh and eat them once they have located the body. The damage inflicted upon a body is more often directly related to the number of crustaceans present rather than to their size. Lobsters and crabs, the larger members of this family, are often responsible for scarring or eating a portion of a body found in water, while their smaller cousins, the crayfish and prawns that are usually found in greater abundance, may effectively reduce a body to a near-skeleton within a week. Still smaller and more numerous are saltwater shrimp that when present in abundance have been known to remove nearly all flesh from a body in less than 12 hours.

In many cases, loss of nearly half the blood volume of a body may be experienced as it is removed from the water. This is due to anthropophagy-related tissue damage.

The attack on a body by crustaceans is not a random occurrence. The first site of their attack is usually at the site of an open wound. Here there is little resistance, and the softer tissues underlying the skin are exposed and easily accessed. Where there is an open, exposed wound, it will be quickly enlarged by crustaceans. Wounds caused by gunshots, penetration (stabbing), propeller injuries, or other damage will speed the onset of anthropophagy. If there are sufficient crustaceans in the vicinity, the soft tissue wounds may be obliterated.

In the absence of any open wound, crustaceans usually prefer to attack softer, fleshy tissues. These softer tissues include (in order of preference) the lips, eyelids, ears, and nose. In most cases, only areas of exposed flesh will be chosen; clothing, for some unclear reason, serves to provide a rather effective barrier for at least the first 24 hours. Where the body is well-exposed, as in the case of a swimmer wearing only a bathing suit, any location may be subject to attack, but the areas of soft tissue or thin skin will usually be chosen first.

Anthropophagy is not limited to saltwater or brackish water. While it is safe to expect large varieties and quantities of marine life in the oceans, scavengers may be found in most bodies of water, from swamps to clear mountain streams.

FISH

When considering the material most commonly used for bait by sport fishermen, it should not be too surprising to discover that fish are often responsible for anthropophagy. Most fish are predators or scavengers, with the majority fulfilling both roles, depending on the availability of the food source.

When the topic of fish consuming a carcass or human body is discussed, the topic of piranha often surfaces. While piranha are aggressive feeders, their reputation for being able to strip a carcass to the bone in a matter of a few minutes is overstated. Piranha were found exclusively in South America until recent years but have been found sporadically in various bodies of water in the southeastern United States in the past decade. Except in South America, which is their homeland, there has been no documentation of this fish aggressively feeding on a human body.

With this in mind, however, it should be clearly understood that almost any freshwater or saltwater fish is capable of feeding on human remains. Larger adult fish such as cod, pike, bass, pickerel, salmon, and even catfish, which are usually classed as predators, are rarely responsible for anthropophagy. There are exceptions to even this rule, however. The juvenile counterparts of these fish are often responsible for feeding on human remains.

Along with these immature fish, there are countless species of smaller fish inhabiting freshwater and saltwater areas that will aggressively seek out and feed upon any carcass in their vicinity. Bearing this in mind, it should be noted, however, that damage caused by fish is usually minor and often noticed only by the trained observer. This feeding on human remains by these small fish is usually documented as incidental or minor in nearly all cases involving body recoveries from freshwater. While the removal of certain parts of the face and hands may not seem a minor occurrence, it is indeed small compared to the damage and disfigurement inflicted by crustaceans.

STICKLEBACKS

In most countries of the world, there exists a small fish commonly referred to as the stickleback. The stickleback varies in length from 1-4 inches (2.4-10 cm) and usually has one to five spines protruding from its dorsal (top) fin and one to three spines protruding from its pectoral (side) fins. Its presence has been documented in tens of thousands of lakes and rivers, and in countries and climates from subarctic to equatorial.

The stickleback is a fish that survives largely because its feeding habits range from carnivorous to vegetarian and from scavenging to predatory. Its habitat is extensive, but it seems to prefer areas of high vegetation.

Within hours of a drowning, schools of hundreds of these small fish have been seen actively attempting to feed on a human body. Their attempts seem to be rather futile, perhaps because of their size, but they provide a clue to the public safety diver as to the location of the body — especially in lakes whose bottom is covered with a thick carpet of vegetation. Vegetation such as elodea and millfoil is often responsible for obscuring the body of a drowning victim. However, within this lush growth is often found a large population of sticklebacks. Schools of sticklebacks swimming near the bottom should always be investigated.

SEA LICE

Sea lice is a generic or common term given to a large group of crustaceans that are found primarily in saltwater or brackish water. They are usually one-quarter to 1 inch (6-25 mm) in length, and while they may be seen free-swimming, their usual mode of transportation is via a series of short, closely-spaced legs located on the underside of their bodies. Their color varies from a cream-white to a mid-brown, depending on their location and subspecies.

Sea lice, which are almost never seen by most divers, seem to congregate in groups of thousands within 24 hours in the presence of food.

Sea lice are voracious feeders and will enter the body through wounds or any available orifice. Once inside the body, they will continue to feed uninterrupted. Sea lice, unlike shrimps and other crustaceans who primarily feed on the skin, will also enter a body cavity through any opening of opportunity. These openings include the mouth, nose, ears, eyes (forcing their way around the

A school of three-spined sticklebacks were noticed congregating over a depression in the millfoil (freshwater vegetation). The body of the drowning victim was discovered deep in the weeds upon careful inspection. The swimmer's body was completely covered by the vegetation, but the depression in the millfoil and the school of the several hundred sticklebacks attracted the attention of the public safety divers. The body had been on the bottom less than two hours.

eyeball), anus, etc. Once within the body, they will continue to feed until the source of nourishment is consumed or they are disturbed.

Sea Lice: A Word of Caution

Recovering a body that is blanketed by these invertebrates is not a pleasant task. Upon initial movement of the body, the sea lice residing on its surface will leave quickly, forming a cloudlike configuration around the divers and the victim. To the novice diver, this could present a rather unique experience that could easily lead to a panic response. In addition, it is not uncommon for the smaller varieties of these invertebrates to quickly tuck themselves under a diver's neoprene hood, gloves, or bathing suit. While this is not a serious matter, the inexperienced public safety diver will certainly not appreciate their presence.

These small crustaceans are harmless if removed within a half hour, but even so, their presence on a diver is not to be taken lightly. In short, their infestation on a diver, no matter how slight, could cause undue stress. All public safety divers involved in body recovery where sea lice are present will do well to acquaint themselves with these invertebrates. Wearing an encapsulated drysuit can prevent this additional stress on public safety divers and increase diver safety.

Sea Lice: Victim Identification

Whenever it is necessary to have the victim of a drowning identified by friends or family, it is a trying time for all involved. A special word of caution is appropriate where an infestation of sea lice is present. It is not uncommon for sea lice to be seen exiting a body for up to several hours after its recovery and removal to a hospital or morgue. Invariably this presents a problem should this occur during the victim-identification phase of the investigation. For the next of kin, family or friends to witness these invertebrates making their exit from the nose, mouth, ears or eyes of their loved one is inexcusable.

It is the responsibility of the public safety diver to alert the police and the hospital staff of the possibility of such an occurrence, and it is then the responsibility of the hospital staff to ensure such a macabre scene is never witnessed by the family. There are many ways to temporarily block these orifices, and it is clearly the task of the appropriate hospital employee (doctor, pathologist, nurse, etc.) to ensure that this simple task is done. The role of the public safety diver here is to educate these professionals and ask for assistance, stressing the fact that this is a humanitarian act that is of great importance.

ALLIGATORS

Although alligators are not found widely distributed throughout the world or even the continental United States, no discussion of anthropophagy would be complete without at least mentioning them.

Alligators, although not usually thought of as carrion eaters, are that. In most cases alligators prefer the role of predator, but they are certainly capable of scavenging. Their diet consists of other small reptiles, birds, or mammals. On occasion they are cannibalistic. In general terms, the alligator is considered more of a nuisance than a danger to public safety divers. Their ability to remove a body following a drowning, however, must be accepted, and while it has rarely been documented, alligators will remove a body as a viable food source.

The feeding habit of larger reptiles is unique. After killing or locating their food source they usually (if it is fresh) transport it to their den.

This den is usually nothing more than a recess under a bank. Later, as their appetite demands, the stash is removed and eaten. The alligator, when it cannot swallow its catch whole, will take the prey in its mouth and shake vigorously until it becomes dismembered and the smaller pieces can be swallowed whole. Contrary to popular belief, the alligator does not prefer rotten carrion, but often its prey must be at least partially decomposed before it can be reduced to portions small enough to swallow.

Alligators may indeed be responsible for some unsuccessful searches for drowned and/or missing persons. Because they remove and hide their victims, or finds, it may never truly be known just what role alligators play in this area.

Aside from anthropophagy, the main concern with these reptiles should be one of diver safety. Obviously, to actively search under river banks or among tree roots in swamps for a body that may have been placed there by an alligator would be foolhardy. While this may seem obvious, it is a practice that has been done and should not be repeated.

MAGGOTS

Maggots are a very important form of anthropophage that possesses the ability to provide specific clues as to the body's history after death. These clues include (a) the postmortem interval, (b) general history, and (c) original location of the body, if it has been moved.

Maggots — What Are They?

Maggots are not spontaneously generated. They do not appear in or on a body after death as a result of an organism the body contained during life. Maggots form as a result of invasion by insects. Most maggots are the larval form of common flies. Their size, shape, and color depends on their species.

The Life Cycle of a Maggot

To understand the significance of observing, recovering, and taking samples of maggots, it is first necessary to understand how they form, live, and ultimately disappear.

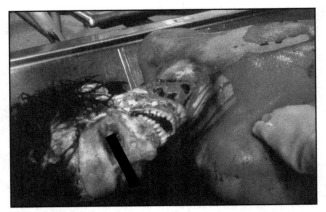

Damage done by sea lice becomes obvious during autopsy. In this case, a loosely fitting collar on the wetsuit allowed these crustaceans to begin at the throat, quickly burrowing into the neck. Sea lice were found deep in the plural cavity and elsewhere.

Blood loss due to anthropophagy can be quite dramatic. When removing the victim's body from the water, take care to minimize the loss. If anthropophagy was noted around the throat of the victim, an attempt to remove the body vertically (head up) from the water — not feet first as was done in this case.

When the victim of a scuba fatality was discovered in 110 feet of seawater, he was blanketed by sea lice. He is lying on his back; his white diver's light, steel tank, and orange snorkel tip are clearly visible, but most of his body is covered with the small crustaceans.

Most flies lay eggs. The site on which they choose to lay their eggs is usually an open wound or orifice of the body (in air). While most species prefer a site where the skin has been broken, sites such as under the eyelids or between the lips are also commonly chosen.

The length of the postmortem interval before flies will begin to lay eggs in or on a body is dependent on the temperature and the accessibility of the body to flies. Where the weather is warm and insects are active and plentiful, flies may begin to lay their eggs within a few minutes after death. Certain species of flies do not lay eggs. These varieties bear live larvae, and these small maggots (larvae) may be

seen moving within an hour after death. Most flies, however, lay eggs that may resemble a collection of yellowish-white specks. Larger flies may lay eggs in sufficient quantities to leave an appearance of grated cheese, deposited in open wounds or around the eyes or lips of the victim.

The time required for these eggs to hatch depends chiefly on the species of fly and the temperature. The majority of eggs, however, hatch in approximately 24 hours. The tiny crawling larvae that leave these eggs are commonly called maggots. These maggots will immediately begin to feed on the body and will grow in size as they do so.

Depending mainly on the temperature, the life cycle of a maggot (fly) is usually a process in which it continues to feed and grow. As these maggots continue their growth, they pass through specific stages referred to as instars. Most maggots will pass through four specific instars as they grow toward adulthood, although some species have up to eight and others only three. Some species pass the first two instars in the egg. In any case, an entomologist may be able to estimate the time since eggs were deposited from the instars of maggots found on the body.

As a maggot approaches maturity, it encases itself in a hard shell. This brownish shell is referred to as a pupa. While inside this shell, the maggot continues to develop and change shape, a process referred to as metamorphosis. During this phase of its development, the maggot is said to pupate. This period of pupation is approximately one week. At the end of this stage of life, the shell is broken, and an adult fly emerges. This life cycle is similar to that of a butterfly.

Immediately after the adult fly emerges from its shell, it will remain quiet for up to several hours while it dries and its wings become functional. The adult fly then takes to the air in search of food and a mate. Flies are capable of breeding almost immediately.

When the eggs of a female are fertilized, they will remain inside her until the larvae are almost fully formed (within the eggs). When ready, she will deposit her eggs on a suitable food source, and the well-formed larvae will hatch, usually within 24 hours. The cycle is complete.

Importance of Maggots
When maggots are seen on a body, their presence indicates that the body has been exposed to air for at least a brief period of time after death. A drowning victim who has never refloated, no matter how badly decomposed the body is or how long it has been submerged, will never display maggots. If maggots are seen on or in a body that has been recovered underwater, then (a) the body must have been exposed to the air after death and before submergence, or (b) the body floated due to internal gas formation. While it was at the surface, fly eggs were deposited. The gases subsequently escaped

from the body and allowed it to sink, taking the maggots with it.

Caution: Flies will lay their eggs in open wounds of living animals and people. In cases where hygiene has been so poor as to allow this to happen, it is possible that maggots may be seen at the site of an open wound. Their number will most assuredly be small, however, and their presence restricted to the site of the wound.

Maggots — What Do They Tell Us?
A trained entomologist, given sufficient information, may be able to approximate a postmortem interval if he can be supplied with an adequate sampling of maggots. This is especially true if no maggot has reached maturity and taken flight as a fly. Where there are empty pupa casings left behind, it may not be possible to determine whether a maggot present on the body is a first-, second-, or third-, etc., generation maggot.

If enough samples are taken and it can be determined that no maggot has matured into a fly at the time the samples were taken, an entomologist may be able to study an adequate sampling of maggots and, by determining the age of the oldest instar, approximate a postmortem interval. This will depend on the ability of the investigator to collect a large enough sampling of maggots. The entomologist must have enough instars to sort and identify the various species since they develop at different rates.

Maggots — Collecting Specimens
To determine the postmortem interval, an effort must be made to collect a large sampling of maggots, including the oldest/largest specimens. As soon as possible after locating the body, a quantity should be collected and preserved. Their growth — hence the postmortem clock — will be stopped when they are dropped into the preserving fluid.

A second quantity of maggots should be collected and kept alive. These should then be taken to an entomologist as quickly as possible. It will be the task of the entomologist, who will subsequently study these collected instars, to use his expertise in exposing the specimens to similar temperature and environmental conditions while providing them with a suitable food source. The entomologist faced with this task may have to allow these larvae to pass through various instars (stages) of their larval growth to ultimate adulthood to determine the exact subspecies of fly. In most cases, the winged adult must be examined.

It is interesting to note that there are hundreds of species of flies and almost countless subspecies in the world. Many subspecies of fly are varieties that are native to specific (and sometimes small) geographical areas. The discovery in a body of a collection of maggots/flies not commonly found in that geographical area may give rise to the suspicion that death occurred at a site other than where the body was recovered.

In Canada and the continental United States there are many rivers and river systems whose length is sufficient to transport a body several hundred miles (hundreds of kilometers). In cases where this has occurred, specific subspecies of maggots found on a body that has floated a considerable distance in a river system may indicate the origin of the body as being several hundred miles (hundreds of kilometers) away. This investigative lead could be invaluable. A similar transportation of a floating body could also be accomplished in a large lake or the ocean. Correct identification of the maggots could assist in locating the body's origin and hastening a criminal investigation while other evidence may still be relatively fresh.

THE SCREWWORM FLY

The screwworm fly deserves special attention for two reasons. First, this member of the insect family has been virtually eradicated in North America through the introduction of sterile colonies. Although it has not surfaced in North America for several years, its emergence in Europe has recently been cause for concern. Undoubtedly, if not controlled at the source, it will once again make its presence known throughout the world.

The screwworm fly is also referred to as *Cochliomyia hominivorax* (man-eater). It is an aggressive insect. The female of this species is attracted to the smell of blood (human or animal) and will commonly lay her eggs in an open wound of a living host. This insect has decimated the agricultural industry by infiltrating whole herds of livestock. It is also a threat to human

health because of its ability to quickly deposit its fertile eggs on a living human host.

When found on a drowning victim, the screwworm-fly maggots may have been deposited before death (antemortem), and their presence could very well be discovered on (or in) a drowning victim who was never exposed to the air postmortem. The screwworm fly is a dark blue-green insect with bright orange eyes. It is slightly larger than a common housefly. The female often attaches herself to the host and lays her eggs immediately. In approximately 24 hours, the eggs hatch and the young maggots begin to bore deep into the host's flesh.

OTHER INSECTS AND LARVAE

Flies are not the only insect to dwell in and on a human body after death. A complete list of insects would number in the thousands. It is indeed important that a sufficient sampling of all life forms be gathered and preserved so that a qualified entomologist can begin to determine a minimum postmortem interval.

Even after all soft tissues have been consumed and only skeletal remains are discovered, a group of insects referred to as carrion beetles may still be found on the body and in the immediate vicinity. On occasion entomologists are able to identify these beetles and estimate a minimum number of months since death.

SAMPLING AND COLLECTING: MAGGOTS AND INSECTS

If a sampling of maggots or insects is to be taken, they must be properly preserved so that they retain their size, shape and color for scrutiny by a qualified entomologist. Such samplings are indeed evidence, and like all evidence they must be preserved in their original condition.

It must be emphasized that a large number (often more than a hundred) of maggots or insects should be collected. This collection should contain chiefly the largest (and oldest) maggots present as well as the smallest, along with any eggs or egg cases. In short, the collection cannot be too large. Of course, practically the number of maggots taken in a sample may be influenced by the degree of infestation. If only a few small maggots can be located, then they may afford an adequate sample. If there are a few hundred maggots present, a sampling consisting of a hundred or more will likely suffice. In cases where the infestation involves tens of thousands, a sample of a minimum of 200 would be appropriate. In short, when collecting maggots (or any other insect larvae), it is always advisable to err on the side of abundance.

In addition to the collection and preservation of maggots, a large sampling of live maggots must be collected. This collection must be kept alive so that their growth rate will yield the length of time they may have resided on the body; the conditions they are found in must also be duplicated in the laboratory. In some cases, it will be requested that a quantity of the actual flesh these instars were feeding on be supplied along with the living colony. This will ensure that their nutritional requirements are duplicated. The taking of flesh from a cadaver, however, is not an activity that should be considered without permission from the appropriate authority. The pathologist, coroner, and/or other authority may be required to make the ultimate decision.

When collecting these larvae, it is imperative to record the date, time, temperature, and the specific environmental conditions found in the immediate vicinity. Day and night temperatures, moistness of the soil, exposure to direct sunlight, even a sample of soil and vegetation found immediately beside the body must all be collected. At the very least, the collection and preservation of maggots should be seriously considered by the investigative agency.

If no other means of preservation is available, the maggots may be temporarily preserved by simply dropping them in boiling water for two to three minutes. An ideal preserving fluid consists of 85-90 percent alcohol, 10-15 percent formalin, and 5 percent glycerin. These chemicals are readily available at most drug stores. Specimens ranging from sea lice to maggots, when preserved in this fluid, retain their original shape, color, and size indefinitely.

Conclusion — Why Collect Maggots?

1. The presence of maggots indicates that the body was exposed to air after death.

2. The collection and study of maggots may assist in determining a minimum postmortem interval.

3. In cases where the body has been transported many miles/kilometers by a river, ocean, or large lake, the determination of the precise subspecies of insect may provide an investigative clue to the geographical origin of the body.

ANTHROPOPHAGY — REASON FOR A SPEEDY BODY RECOVERY

Aside from the obvious fact that anthropophagy may render an otherwise presentable body disfigured, even beyond recognition, there is a second reason why a professional public safety dive team should be selected for a speedy body recovery whenever possible. Anthropophagy usually begins at sites of wounds and quickly alters the appearance of any wound that was inflicted upon the victim.

A well-trained, observant and efficient public safety investigative diver is a mandatory requirement for the recovery of any body — regardless of whether death was accidental or homicidal. The speed at which a body should be recovered, of course, is somewhat slower and more cautious than if the dive team is responding in a rescue mode, but in areas known to be frequented by large populations of scavengers, a degree of urgency should be encouraged. A speedy body recovery not only will result in a more efficient, accurate autopsy but also may make victim identification less traumatic for family and friends. As trained professionals, we must remember that the people who are left behind are also victims.

Decomposition

Without doubt, the most strikingly distasteful aspect of body recovery presents itself when putrefied human remains must be recovered and examined by the public safety diver. Like many other postmortem changes, putrefaction, when it is present, cannot be ignored. The presence (or absence) of putrefaction may give valuable clues as to the time of death (postmortem interval). In addition, the mere presence of decomposition may hide wounds that would easily explain the original cause of death.

The process of postmortem decomposition is based on two processes: autolysis and putrefaction. While these two processes may work simultaneously within the body, their mechanisms of action are quite different. When a body is in an advanced state of decay, it is often said to be in a state of decomposition. More accurately, the body would be described as exhibiting an advanced state of autolysis or putrefaction (or both). Decomposition must be classified as either autolysis or putrefaction.

AUTOLYSIS

The *Merriam-Webster Medical Dictionary* defines autolysis as "breakdown of all or part of a cell or tissue by self-produced enzymes — called also self-digestion."

In a human body, autolysis occurs after death when digestive enzymes continue to work within the body without the normal controls present during life. Normally, these digestive enzymes begin their work in the stomach where hydrochloric acid is introduced. Adequate secretion of mucus by the stomach lining protects the stomach from damage by this powerful acid. As the hydrochloric acid and other enzymes break down food, it is passed on to the small intestine where further enzymes are introduced. During the normal process of digestion, these acids and enzymes react to break down complex-carbohydrate and protein molecules to simpler molecules, which can be absorbed into the bloodstream and provide nourishment and energy to the living body.

During life, our gastrointestinal (digestive) tract actively produces mucus, which protects the living tissue from destruction. After death, in the absence of this defense, these acids and enzymes will continue to react, and eventually they will break down the tissues of the gastrointestinal tract. When this occurs, perforation of the stomach, lower esophagus and intestine will occur. Once these digestive enzymes seep into the abdomen or pleural (lung) cavity, their reaction with the internal organs of the body may continue unabated. Simply described, the body begins to digest itself.

Often these enzymes seep into the pleural cavity within hours. When these enzymes react with the lungs, the lung tissue is destroyed quickly. This action explains why a drowning victim

This victim of a diving accident was recovered within 24 hours. He was a breath-hold diver, yet upon recovery his mask was seen to be filled with blood. An autopsy confirmed the cause of death as drowning and the origin of the blood as autolysis.

The body of this ice-diving victim exhibited a noticeable blood loss immediately upon removal from the water. An autopsy revealed that the lung damage was a combination of autolysis and hemorrhage due to the forceful agonal-gasp phase common in many drownings. The cause of death in this case was asphyxia due to drowning.

recovered within 24 hours may display virtually no evidence of decomposition, yet when the body is removed from the water, a bubbly, malodorous, sanguinous (blood-stained) fluid exudes from the mouth and nostrils. In cases where autolysis is more advanced, this fluid will take on a more watery consistency with a light brown-green color. Its odor is quite characteristic and intense.

This dissolving or autolysis of the lungs is called pulmonary autolysis. This process is often hastened by stomach contents (gastric juices and acids), which are breathed into the lungs during the convulsive-vomiting and agonal-gasp phases of drowning. Because of the digestive juices that may be drawn into the lungs during the agonal-gasp phase, pulmonary

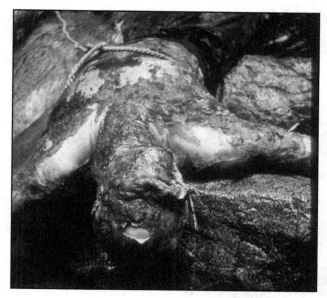

The skin in an advanced state of putrefaction softens and becomes easily detached from its supporting structures. Skin that has not been disturbed may exhibit a patchwork of colors and textures. Despite this, the structure of the body remains firm, allowing for ease of handling. This subject had been found after being in 55°F water for 28 days.

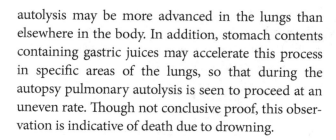

Prolonged submergence in water often results in partial or total detachment of the epidermis (skin layer) of the hands (palmar region) and feet (plantar region). Great care must be taken when handling a body since ultimate identification may depend on fingerprints being taken that are suitable for comparison.

autolysis may be more advanced in the lungs than elsewhere in the body. In addition, stomach contents containing gastric juices may accelerate this process in specific areas of the lungs, so that during the autopsy pulmonary autolysis is seen to proceed at an uneven rate. Though not conclusive proof, this observation is indicative of death due to drowning.

Summary

Autolysis is a dissolving of the tissues in the pleural (lung) cavity and abdomen due to the seepage of digestive juices. The onset of autolysis usually precedes putrefaction. Pulmonary autolysis is primarily responsible for the malodorous, brownish-green, blood-stained fluid often seen seeping from the nose and mouth of a drowning victim whose body is recovered before putrefaction has begun. Autolysis is not putrefaction.

PUTREFACTION

Putrefaction is the major component of the process commonly called decomposition. It is a bacterial action that proceeds at an accelerated rate postmortem, when the natural defenses of the living body cease to hold bacteria in check.

A simple analogy of putrefaction would be to compare bacteria in a living body with prisoners in a jail. In life, the prisoners (bacteria) are kept alive but under control. After death has occurred, the iron bars and prison guards disappear, letting out the bacteria (prisoners) to wreak havoc and eventually consume the system that maintained and controlled them while it was alive.

The bacteria present after death multiply quickly, and when sufficient numbers exist, they possess the capability to literally digest the soft tissues of the body, reducing them to a fluid consistency. A byproduct of this process is a complex mixture of foul-smelling gases.

After remaining in clean, mountain river water (62°F/16.7C) for 14 days, the body of this victim shows very little putrefaction on the surface of his skin and is in remarkably good condition.

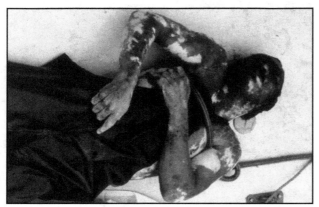

The body of this victim was recovered after only 24 hours of submergence. The water temperature was 60°F (15.5C). Advanced putrefaction on the skin was influenced by a nearby sewage outfall, where ambient bacteria rapidly accelerated skin discoloration. Divers operating in these conditions must take appropriate precautions.

Putrefaction — A Closer Look

Putrefaction is the most visible representative of death encountered by public safety divers. Even so, when asked to describe it in detail, it is usually only spoken of in general and vague terms.

When reporting on the involvement and degree of putrefaction, specific details should be included. For this reason, it will help to understand exactly what is happening to the body as putrefaction progresses.

As countless billions of various types of bacteria multiply within the body, gases are formed in all areas of the body. Locations deep within various organs as well as within the circulatory system may be sites of bacteria-generated gas.

When gas production occurs within the body's blood vessels as well as soft tissues (e.g., muscle tissue, organs, etc.), the body becomes distended and appears swollen. Skin color changes through blue, green, purple, brown, and eventually to black. The eyes and tongue may be forced to protrude well out from their supporting structures. Because of increased pressure in the viscera (abdomen), there may even be gross protrusion of the abdominal organs through the vaginal and rectal openings.

The skin in an advanced state of putrefaction softens and becomes easily detached from its supporting structures. If roughly handled, large areas of skin may easily be removed from the body, even though the arms and legs are still firmly attached to the trunk.

The skin may be evenly discolored, or it may exhibit a patchwork of colors, depending on the varieties of bacteria present and the degree of putrefaction.

Since most pigment is contained within the skin, as this layer of skin becomes dislodged through rough handling, the underlying pink musculature of the arms or legs may be revealed.

Caution: Skin slippage is present in all cases of advanced putrefaction. Where the identity of the body has not been determined, the victim's body should be treated very carefully.

Fingerprint examination is often used for positive identification, and when putrefaction is well-established, the entire epidermis (skin layer) of the hands may become detached in a glovelike configuration. Areas of skin still retaining a usable fingerprint may become detached and lost unless the body is treated with great care (see "Fingerprinting the Deceased").

Factors Affecting the Speed of Putrefaction

The onset of decomposition and the speed at which it progresses is subject to several variables. Each one of these variables may itself either retard or hasten the speed of putrefaction many times. Determining the post-mortem interval using simple observations of the physical state of the body is, at best, a highly risky exercise.

Following are some of the factors affecting the rate of onset and the progression of putrefaction:

Temperature

Temperature has the greatest influence of all variables on the rate of putrefaction. In cold climates where the temperature is near or below freezing, a body may be preserved indefinitely. In instances where a drowning occurred in water that was 34°F (1.1°C) or colder, a body may remain in relatively good condition throughout the winter months. However, once the body has been recovered and removed from the cold water into a warmer air environment, putrefaction may proceed at an alarmingly accelerated rate.

Conversely, drowning victims who have been recovered in water warmer than 75°F (23.9°C) after only 48 hours have been found to be in an advanced state of putrefaction. Indeed, many bodies have been recovered in an advanced state of putrefaction within 12 to 18 hours after drowning in water where the temperature was measured at 80°F to 85°F (26.7°C to 29.4°C). These bodies have been so disfigured that facial features were no longer identifiable. When this occurs, the hair in the area of the scalp may slip away, and the body may be swollen to nearly twice its normal size.

Obesity

It is a known fact that obesity hastens putrefaction. The exact reason for this is not entirely clear, but it is an important fact to consider when searching for the body of a drowning victim, since putrefaction produces gases. The topic of gas production is tied in with the subject of obesity, and when considering the time for "refloat" of a drowning victim, obesity is a factor that should be considered.

Health

Since putrefaction is the product of bacterial action, the body of a person suffering from any bacterial infection would be prone to exhibit a greater degree of putrefaction in a shorter period of time than would the body of an otherwise similar but healthy accident victim.

Environment

While most putrefaction is caused by residual bacteria already living in the body, bacteria in the surrounding air, water or soil cannot be ignored. Ambient (surrounding) bacteria may play an integral part in the initial appearance and progression of putrefaction.

Putrefaction in a drowning victim would be expected to proceed at a much slower rate in a clean, clear mountain stream than in a stagnant body of water. Stagnant bodies of water harbor their own multitude of ambient bacteria, and it is not surprising that bodies recovered from these waters often exhibit an astonishingly advanced rate of putrefaction. In cases where ambient bacteria abound, putrefaction may well precede any visible bloating or gas formation.

Newborn Infants

When a baby is first born, his digestive tract holds only sterile meconium. With breast milk and other food, in time various forms of bacteria enter and take up residence in the intestine. Such bacteria will be present throughout the life of the individual. As has already been discussed, this bacteria is often the source of putrefaction. Since a newborn (or stillborn) infant may not yet possess these bacteria, decomposition may take place very slowly. In addition to putrefaction being held back, autolysis may also be very insignificant, with the gastric juices virtually missing. In the absence of resident bacteria and digestive juices, there remains almost nothing to begin the processes of autolysis or putrefaction.

While the process of mummification would not apply to the body of a newborn recovered from the water, it is important to understand that if left on land, in the absence of quick decomposition, a newborn presents a very small volume of tissue. This small volume can be dried very quickly. It is not uncommon to recover a fully mummified body if a newborn has been discarded in a warm, well-ventilated, and dry location.

The Decomposition Clock

Due to the great number of variables that affect the onset and rate of decomposition, and the fact that any one variable could easily double or triple the speed of decomposition, the processes of decomposition (autolysis and putrefaction) do not afford the investigator with an accurate clock with which to assess the time of death or postmortem interval.

The chronology of events involved in decomposition, however, could at least assist the investigator in determining to what degree the body is decomposed. The phrases "very decomposed" or "advanced state of decomposition" are subjective evaluations and should be avoided. Instead, the report should be in clear, descriptive, and objective terms.

Decomposition is a transition process that, when complete, returns the body to the elements from which it was originally created. It is a natural and methodical process that can be reported on in a clear, concise manner.

The Chronology of Decomposition

The following sequence of events represents in considerable detail the process of decomposition. This timetable would normally be followed in a temperature range of 68°F to 73°F (20°C to 22.8°C).

Twelve to Twenty-Four Hours

The skin is the first to change color, from normal to light blue to an almost green discoloration. This color change usually occurs in the lower quadrant first, being most noticeable in the area of the lower abdomen, pelvis, and groin.

Twenty-Four to Thirty-Six Hours

The discoloration becomes quite pronounced, and the skin takes on a marbled pattern. The blood is now reacting with hydrogen sulfide, which has been produced within the blood vessels. This creates a characteristic dark green (almost black) discoloration. As putrefaction advances, the blood seeps from the blood vessels, giving the body a general purplish-black color.

Thirty-Six to Forty-Eight Hours

The face and trunk begin to swell noticeably, taking on the characteristic bloated appearance. The eyelids, lips, scrotum, and other sites where skin is loosely attached may become dramatically swollen and bloated. On palpation (feeling or manipulation) of these areas, crepitus (a feeling described as "Rice Krispies") is noticed.

Sixty to Seventy-Two Hours

Putrefaction has now spread to all areas of the body, including the fingers and toes. The entire body has now changed color, and facial features may become unrecognizable.

Four to Seven Days

Hair and nails become loose and are easily removed unless the body is handled very carefully. The skin covering is easily damaged or torn loose. The body should still retain its structural integrity if carefully lifted by both arms and legs.

Pockets of foul-smelling gas usually form under the skin. These gas pockets may easily escape, allowing this putrid-smelling gas to exit the body through any tears in the skin. As well as gases formed through putrefaction, malodorous, colored liquids may escape from the body from natural orifices, wounds, or skin ruptures caused by either rough handling, injury, or putrefaction. When putrefaction is advanced, the skin of the hands (palmar) and feet (plantar) may become easily detached and make subsequent fingerprinting impossible.

Two Weeks and Beyond

In time, all soft tissues of the body (if not dried by heat) will be reduced to a gray, greasy, unrecognizable mass. Eventually only skeletal remains are left behind to indicate the prior presence of a human body. In time, the skeleton will also be consumed.

REFLOTATION INTERVAL (GAS PRODUCTION)

In the past decade much has been written, along with countless hours of research performed, in an attempt to answer the very basic question: How long will it take a body to refloat? Perhaps the chief reason why this question has never been answered satisfactorily is because it is not a simple question. Indeed, the reason why any body (human or otherwise) would refloat after being submerged in water is merely because of gas production within the body.

Understanding this intangible refloat enigma involves understanding the factors influencing gas production within the body.

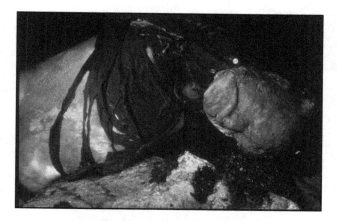

As putrefaction advances, the face and trunk begin to swell. Hair and fingernails become loose, and facial features may become unrecognizable.

Further putrefaction results in the loosening of skin covering. It is easily damaged. Pockets of foul-smelling gas usually form under the skin. The body changes color and may exhibit a variety of color and textures. Internal organs may be reduced to a gray, greasy unrecognizable mass. The body will begin to lose structural integrity and may become dismembered if not carefully handled.

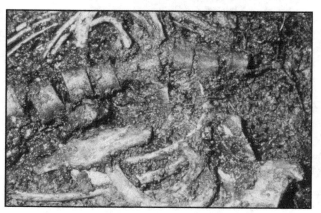

As putrefaction (and anthropophagy) advances, the body may be reduced to the appearance of a skeleton. At least for a while, internal organs will remain in a semiliquid state. Handling a body in this advance state must be done with incredible care.

Eventually all soft tissues disappear, leaving only skeletal remains. The bones that are left may settle into the bottom of the body of water, losing their characteristic human shape. Eventually even the skeleton will be consumed.

Origin and Nature of Refloat Gases

Gases that are produced within a putrefying cadaver are primarily (though not exclusively) carbon dioxide, hydrogen sulfide, ammonia, methane, and sulfur dioxide. These gases are formed in varying quantities and percentages. Rarely will any two putrefying cadavers produce the same gases in the same quantities.

These gases, which are formed as a result of bacterial action, are produced at various sites within the body. As these gases form, buoyancy is created, and the body, in many cases, eventually resurfaces in what is commonly referred to as the reflotation phenomenon. When and how the victim will resurface depends on many factors, including the site of the gas production within the body. Since gas is produced in (chiefly) two different locations, each will be discussed separately.

Primary Flotation

Due to the large quantity and variety of resident bacteria in the digestive tract, gases are constantly being formed as a natural byproduct of normal digestion. This gas production is chiefly dependent on the variety and distribution (as well as the types) of bacteria present. Another major factor in the production of these gastrointestinal gases is the composition of the most recent meal consumed by the individual. Meals high in carbohydrates tend to produce some gases very quickly. In addition, carbonated beverages are warmed as they enter the digestive system, releasing carbon dioxide. Since death does not necessarily slow the production of these gases, they continue to form even after a drowning. Indeed, upon death, certain digestive enzymes that were inhibiting specific bacteria may cease to be produced, resulting in an increased gas production. Primary flotation, then, is caused by gases that are formed in the digestive tract. These gases are formed quickly and radically affect the buoyancy of a drowning victim in the first 24 to 72 hours.

A body that has surfaced due to the formation of these gases in most cases will exhibit a distended abdomen, often appearing somewhat obese. Moving such a body may result in the escape of at least a portion of the contained, foul-smelling gas as well as a loss of buoyancy to the body. In some cases, bodies that have surfaced due to primary flotation

Primary flotation is usually a result of gases produced within the digestive tract. A distended abdomen is characteristic of primary flotation. Putrefactive gases formed during the primary flotation phase may be released through the mouth, nose, and anus during the handling and removal of the body. This individual refloated after less than 48 hours in 68°F water.

Secondary flotation involves gas production in most of the bodies' tissues. It is characterized by a general swelling of the limbs, trunk, face, etc. During primary flotation the body will usually float face down, presenting a very low profile above the water. This victim, who is exhibiting secondary flotation, is very buoyant. During secondary flotation they may float face up or face down due to their buoyancy. The blackened face of this body is a result of a combination of putrefaction and exposure to direct sunlight for several days.

release some of their gases upon arrival at the surface and, having lost their buoyancy, sink. In most cases, they will remain on the bottom until secondary flotation occurs.

Secondary Flotation

Secondary flotation occurs when bacteria present within the body produce gas at sites other than the gastrointestinal tract. In this case, the gases are formed equally throughout the body, not only within actual muscle tissue but also at other sites such as the circulatory system and viscera. When a body refloats due to gas production at these sites, its position/buoyancy on the surface is considered stable. A body that has floated due to secondary flotation is easily recognizable by the appearance of its swollen limbs. Even so, in cases where gas has formed throughout the body, the limbs may or may not be grotesquely swollen or enlarged.

Factors Affecting the Time to Refloat

Over the years much has been written on this topic. Charts have been produced relating water temperature to time to refloat; graphs, calculations, even scientific research using a variety of animals has failed to produce specific workable formulae that could be used to calculate (with any degree of accuracy) actual time to refloat for a drowning victim. The following variables and how they hasten or delay refloat should explain why this will never be an exact science.

Last Meal

Both the time and contents of the last meal will have an effect on reflotation time. Meals high in carbohydrates tend to produce gas fairly quickly, while some types of food actually inhibit the formation of gas. In addition, a meal that has been consumed and is still in the stomach may be less prone to producing gas than one that has passed into the small intestine. At the very least, gases that are formed in the stomach are more quickly released (as they are produced) than gases that form further along in the digestive tract. The variables in this case then become (a) contents of the last meal and (b) time since its consumption. Of course, the quantity eaten will also have an effect.

Temperature

Gases that are formed in a cadaver are formed as a result of bacterial action. Bacterial action is increased by elevated temperatures. It follows then that warm water will result in a greater (quicker) gas production and a shorter submergence time before refloat.

Depth

Most of the gases formed in the putrefactive process have two common properties: They are highly soluble in water and easily compressed. Bearing this in mind, it is easy to understand that depth has a great effect on the time to refloat. At depth, the volume of the gas formed is considerably diminished, and this, coupled with its high solubility in

water (and surrounding tissues), makes depth a true deterrent to reflotation. It is believed (but not proven) that bodies that come to rest in depths greater than 200 feet (60 m) will not refloat. In truth, they may, but such refloats would be very rare indeed. Gases formed at these depths possess not only insufficient volume to effect a flotation, but cold temperatures and increased pressure make them very soluble in the water and body tissues. Some gases may not even form.

Body Mass Index

Body mass index is a term given to the ratio between lean and fat tissue in a body. A drowning victim who possessed a large quantity of body fat would likely refloat very quickly (compared to a lean individual under identical conditions). The reason for this is twofold: First, possessing a large quantity of body fat tends to supply the victim with additional buoyancy even before any gas production. Second, it has been documented that fat bodies putrefy (and produce gas) at a faster rate than lean bodies — although the exact mechanism behind this is not known (possibly heat retention; see "Factors Influencing Algor Mortis"). In short, an obese body is likely to resurface more quickly than a lean one.

Health

The health of an individual antemortem will have a bearing on his time to refloat. All living bodies contain myriads of microorganisms in the digestive tract and elsewhere. It is a large percentage of these microorganisms that ultimately produce the gases that effect flotation. Hence, the quantity and variety of bacteria (microorganisms) present within a living body will determine (to some extent, at least) the quantity and type of gas that will be formed following death.

In short: A rule of thumb states that the healthier an individual is, the more time his body will take to produce sufficient gas to refloat after submergence in water.

Conclusion

Despite the scientific research that is presently being conducted, it is not likely that an accurate scale will ever be developed that will enable the public safety diver to determine time to refloat. The variables are many, and each one has the capability of drastically affecting the outcome. Research currently is being carried out on animals and human cadavers (current research is being conducted using human cadavers in at least two countries), but any conclusions reached will be valid only for the animals (or cadavers) used and apply only to the conditions under which the tests are performed.

In life, the variables are many and the possibilities endless. The public safety diver is cautioned against predicting time to refloat for any specific scenario. At best, the variables should be discussed with the investigating agency and their people allowed to make their own predictions.

SAPONIFICATION: AN ALTERNATIVE TO DECOMPOSITION

Until recent decades, it has always been thought that only cold temperatures could measurably retard the onset or progression of decomposition. Except for mummification (extreme drying of soft body tissues) or man's own intervention, historical reports of bodies being found months or years after death and still in relatively good condition were sporadic and anecdotal.

In the past few years, however, there have been many cases of bodies being recovered, often from aquatic environments, where decomposition was held back by some phenomenon or process. This process is called saponification. Saponification is the process whereby certain soft tissues are said to "saponify," which literally defined means "to make soap."

SAPONIFICATION: THE PROCESS EXPLAINED

Early pioneers to North America would make soap by combining animal fat, ammonia, and caustic lye (a strong alkaline substance). The concoction would be thoroughly mixed and heated. This resulted in the animal fat undergoing a chemical change and ultimately becoming soap, a process called saponification. Saponification in a human body, when conditions are favorable, is a similar procedure, albeit not quite as dramatic.

When a body is left in a moist environment or submerged in water, saponification may occur

and dramatically inhibit further decomposition. The process of saponification begins after decomposition has loosened and even partially removed a layer of skin. The underlying (subcutaneous) fatty layer is exposed. In a warm, moist environment, this fat undergoes a process called hydrolysis, with the formation of free fatty acids. These acids combine with calcium and ammonium ions to form insoluble soaps.

The resultant substance, which is indeed a sort of soap, is manufactured from the layer of fat lying just beneath the skin. This fat layer is also referred to as the adipose (fat) layer. The "soap" rendered from this adipose substance (fat) is called adipocere. Adipocere is a rancid, greasy, whitish/gray, soapy, waxlike substance. It first appears in the subcutaneous tissues in the areas of the trunk and buttocks and eventually spreads to other fat deposits. The formation of adipocere inhibits further decomposition.

Factors Favoring Adipocere Formation
Saponification resulting in the formation of adipocere favors a warm, moist, anaerobic (without oxygen) environment. Fat and water alone do not form adipocere. Putrefactive organisms such as *Clostridium perfringens* are very active in anaerobic environments and play an important role in the formation of adipocere.

Factors Inhibiting Adipocere Formation
The presence of air, oxygen in particular, hastens aerobic ("with oxygen") putrefaction and may inhibit or even totally prevent the eventual formation of adipocere. Since adipocere is formed from fats contained within the body, emaciated bodies seldom saponify to any great extent. Dry environments will also inhibit or prevent saponification.

THE SAPONIFICATION CLOCK
Since the process of saponification and the resultant formation of adipocere may effectively retard further decomposition, they will render the decomposition clock useless. Saponification, however, provides its own clue to the postmortem interval. The presence of adipocere usually indicates a postmortem interval of at least several months duration, although there have been rare instances in which adipocere was present within three and one-half weeks after death.

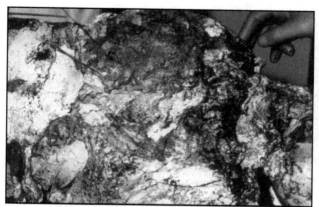

Adipocere appears a gray-white waxy substance. The body of this homicide victim was recovered one year after she was murdered. Her body had traveled a distance of approximately 80 miles down a river and into the ocan where it was eventually found floating. Thin rods have been inserted in the back of her body to designate holes (and trajectory) made by the bullets. In this case, saponification preserved the body of this homicide victim well enough for further evidence to be gathered.

Adipocere formation preserved the body of this homicide victim for one full year. The hole clearly visible in the center of the photograph was made by one of the many bullets fired into the victim by an assailant. An intact lead slug was recovered from the body, suitable for ballistics examination.

It is important to remember that saponification is the process, and adipocere is the product.

Conclusion

Adipocere formation is not a common occurrence. It can drastically interfere with the normal rate of decomposition. In any body recovered from warm water (65°F/18.3°C or warmer), adipocere formation is possible. When adipocere is seen or even suspected, it should be entered in the investigator's report as an important observation. If the conditions that favor the formation of adipocere were not (or never have been) present in the location where the body was found, postmortem movement of the body should be suspected. This is particularly evident where bodies have originally died in a swamplike location or condition, saponified, then later moved as seasonal high waters flooded the area.

Fingerprinting the Body

Fingerprints are used as a means to identify bodies. In some instances, fingerprints are taken merely to confirm the identify. Fingerprint identification is utilized worldwide and is universally accepted as a means of positive identification.

Because of the certainty of fingerprint identification, an unidentified human body should never be committed to the grave until suitable fingerprint impressions have been taken.

One of the last tissues of the body to be affected by decomposition is the skin of the palmar and plantar regions. Fingerprints suitable for identification purposes have been recorded up to several months postmortem. For this reason it should be considered standard procedure to fingerprint a body of either unknown or questionable identity before burial or cremation.

Often fingerprints will be taken only when it is suspected that the victim has his fingerprints on file, either through previous criminal convictions or as a means of security clearance (military or police personnel files, etc.). Fingerprint identification has been used to identify victims of drowning when no such fingerprints were on file. In cases such as this, comparison may be made with latent fingerprint impressions lifted from articles that were known to have been handled by the deceased prior to death.

Why Fingerprint the Deceased?

Fingerprinting the deceased may serve one of three purposes:

1. Positively identifying unknown bodies or confirming the identify of bodies where decomposition has rendered facial characteristics unrecognizable.

2. Identifying the victim in a homicide case. For trial purposes, it may become necessary to prove the identity of the victim.

3. For eliminating the victim's fingerprints that may be located at the scene of a murder. Once the victim's fingerprints have been eliminated, the remaining prints left at the scene are those of others who were present. These others then become suspects.

FINGERPRINTING METHODS
Body in Good Condition
If the body to be fingerprinted is in relatively good condition, the fingers may be inked using an inked roller and prints taken using a standard cadaver spoon. Fingerprint impressions should always be taken laterally from nail edge to nail edge and vertically from fingertip to below the crease of the first joint. After taking each fingerprint impression, it should be studied; if less than perfect, a second or third impression should be taken. It may be necessary to clean the finger and begin again using a lighter or darker ink coat. Each fingerprint impression should contain the appropriate reference as to which hand and finger the impression was taken from (i.e., right/left thumb, forefinger, middle finger, ring finger, or small finger).

If rigor mortis is well established, considerable force may be required to straighten each finger, and the task of fingerprinting may require the use of an assistant. In extreme cases where hands are clenched tightly, it may become necessary to break the fingers or cut the tendons that immobilize the fingers. This should be done only as a last resort and only with prior permission of the coroner and the investigating agency.

Body in Good Condition — Finger Wrinkled (Wauschaut)
One of the simplest techniques to fingerprint a cadaver where prolonged submergence has wrinkled the fingers is to inject air (using a hypodermic needle) into the bulbous portion of the finger by inserting the needle just below the first crease. If a small-gauge needle (28 gauge or smaller) is utilized there will be very little leakage of air during the fingerprinting

procedure. A finger, once inflated, usually presents an excellent fingerprint impression.

Another technique for causing the print surface to swell is to wrap a thin cord around the base of the finger and continue wrapping the finger until the cord has reached the crease just below the first joint. This technique has a tendency to swell the bulbous portion of the finger, making it easier to obtain a reasonably good fingerprint impression. While this technique is certainly less invasive, it is not usually considered as efficient as the injection method.

Perhaps one of the simplest and most effective techniques is merely to apply black fingerprint powder to the fingers with a brush. Using a white/translucent medical tape (Blenderm), lift the fingerprint from the cadaver. Once this tape is applied to a transparent backing, it may be viewed through the backing, as it will now appear laterally correct.

Fingerprinting the Decomposed Body
Obtaining acceptable fingerprints from a body recovered after prolonged submergence (several weeks) in water presents its own unique problems. In cases of prolonged submergence, it is often found that the surface skin (epidermis) is in the process of becoming detached from the underlying tissue (dermis). When this is observed and is seen to be well-advanced, the epidermis may be carefully removed using a sharp scalpel and a great deal of caution. The layer of skin (fingerprint) may then be soaked in lactic phenol solution for approximately one week. After soaking in this solution, it may be removed, wiped dry, and rolled on ink (using your own finger for support), then imprinted on paper.

Soaking the epidermis in lactic phenol for one week (or longer if necessary) will cause the tissue to develop a rubbery, latexlike condition that is impervious to further decomposition.

If the epidermis of the palmar region is missing, it may be possible to obtain a fingerprint impression from the underlying dermis by merely soaking the finger in lactic phenol for several days. In these cases, the fingers must be removed.

Lactic Phenol

A solution known as lactic phenol may be utilized at all times when it becomes necessary to soften finger/skin tissue to render it pliable. When prolonged submergence has taken place, the fingers of the victim are often very wrinkled. The skin is also quite hard. Alcohol or formaldehyde should never be used, since they will only accelerate the hardening of the tissue. Alcohol or formaldehyde is recommended only if the finger/tissue is to be stored or preserved indefinitely, after a suitable fingerprint impression is taken. Lactic phenol excels as a restorative fluid, but is not recommended for long-term preservation. The lactic phenol solution is made from the following ingredients:

1 part distilled water

1 part glycerin

1 part lactic acid

1 part phenol (carbolic acid)

Approximately 16 ounces (475 cc) of lactic phenol is sufficient for the restoration of 10 fingers. Lactic phenol does not deteriorate and may be kept on hand indefinitely.

Fingerprinting Tips

1. When fingerprinting a body, ensure the best possible prints are taken the first time. If possible, have a fingerprint technician inspect and approve the impressions you have taken. There may not be another opportunity.

2. When removing either tissue or whole fingers, it is important to note and record each finger or finger fragment, identifying it as a specific digit and hand.

3. If the fingers of a cadaver are dirty, they may be easily cleaned with wood alcohol (methanol or methyl hydrate). If the tissue is decomposed, great care must be taken to prevent tissue damage.

4. Tissue that has become very distorted and hardened may take weeks of softening in lactic phenol. It should be examined at six-day intervals.

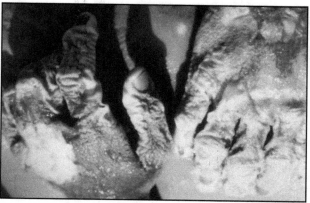

Obtaining fingerprints from cadavers exhibiting wauschaut (washerwoman's hands) presents its own unique problem. This condition has no great significance in determining cause of death, only the submergence of the body in water. In this case, decomposition has not advanced enough to interfere with the taking of fingerprints, but the severe wrinkling of the palmar region of the hands would seem to present an insurmountable problem. Air introduced into the fingertip through a hypodermic needle just below the first crease will cause this portion to swell enough to obtain a suitable surface for fingerprinting.

5. Practice. Taking good fingerprint impressions is not a simple task. The greatest practice possible would be easy compared to solving the problems caused by inadequate or incomplete fingerprint impressions taken during a homicide investigation. It is easy to simulate wauschaut merely by soaking your own hands in warm water for 20 to 45 minutes then attempting to take an adequate fingerprint impression.

Caution: When working with cadavers, adequate protection is advised for your own health. A minimum safeguard can be provided by using rubber gloves. Avoid purchasing economy/disposable gloves. Adequate protection can only be provided by using rubber surgical gloves or patient examination gloves. These are available at all hospitals and morgues. Other protection provided by the use of cap, gown, and mask is rarely needed. In all cases, however, thorough washing with a good antiseptic soap immediately after working with or around cadavers should be considered standard procedure.

Gloves should be disposed of according to local health regulations, and all instruments (cadaver spoon, etc.) should be disinfected by boiling in water (to which a few drops of carbolic acid have been added) for approximately 15 minutes.

5 RECOVERING SKELETAL REMAINS

The recovery of human skeletal remains presents a unique challenge to the public safety diver. Despite the many precautions and specialized conditions that must be given to this specialized recovery, many dive teams have succumbed to the temptation of collecting merely what was convenient, easily retrievable or deemed to be significant. The careful and painstakingly slow procedure to be followed in the recovery of human skeletal remains cannot be circumvented. The reason is simply this:

In the routine recovery of a human body, all clues as to the identity of the deceased and the cause and time of death may be held firmly intact by the original body bag — the skin. When skeletal remains are to be removed, there is no guarantee that all evidence is easily visible or even in close proximity to the main portion of the skeleton. The remains to be recovered may be scattered over many square feet (meters) of lake, river, or ocean bottom, and the smallest of bones may be obscured by bottom vegetation or silt.

The collecting of skeletal remains underwater then becomes a much more difficult and demanding task than it would be on land.

SKELETAL REMAINS — DEFINITION

Normally, when the term "skeletal remains" is used, images of a clean white skeleton are brought to mind. In reality, skeletal remains are more often partially covered by clothing and flesh. The resulting configuration becomes not only confusing but also incredibly distasteful to the public safety diver, who must personally handle each piece to ensure an efficient, intact recovery.

The term "skeletal remains" could then apply whenever human remains are recovered in which the skeleton (or portions of the skeletal structure) are either partially or wholly exposed. When this

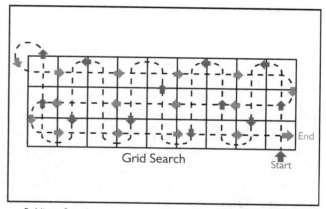

Public safety divers can use a grid search to thoroughly comb an area for small pieces of evidence. As illustrated above, an area is searched by first moving in one direction. The area is then searched perpendicular to the first pass, creating a grid pattern.

occurs, there exists the likelihood of loss of part of the remains during the recovery and subsequent transport to the surface.

The importance of recovering the entire remains cannot be overstated. Since much of the flesh has already been removed, there remains little physical evidence for the pathologist to utilize in determining the who, what, where, when, and how of the death investigation. The use of a grid search is often helpful to the investigator who must locate very small items, such as bone fragments. With less physical evidence present for examination, the loss of or damage to even a small portion of the skeletal remains could mean the loss of a critical piece of evidence, leaving one of the most important questions unanswered — who is it?

Precautions

In the recovering of skeletal remains, it is imperative that everything in the immediate vicinity be recovered. Since much of the evidence may be minute, disguised, or concealed, there are specific precautions and procedures that should be followed.

A public safety dive team in their preparations should realize that this is a unique type of recovery that must be done slowly and carefully. Speed is not important. For this reason, predive preparation is not optional — it is mandatory.

Specialized Equipment

The most important piece of equipment to be used in the recovery of skeletal remains is the container. If the remains are held together by connective tissue (tendons, flesh, etc.), then a specially-designed body bag may be used. Body bags with mesh bottoms are ideal for this purpose, but the mesh should not be coarse enough to allow even the smallest of bones to escape. Full zipper closure is essential. Ideally, the body bag should offer the capability of a horizontal controlled lift and have attachment points for a slow, careful lift from the water. This container should allow for drainage of water and fluids while retaining the contents intact and in order.

Marking the Location

Prior to beginning the removal process, the location of the remains should be marked. Surface markers or buoys should be used to allow the dive team to return to the site for a detailed grid search after the bulk of the remains have been removed. This marker should be carefully anchored close to the location with a suitable heavy weight that will resist movement by wind, waves, currents, or people.

Permanent underwater markers should also be installed to mark the exact location of the remains. Since it may be necessary to return to this site many months later for further investigation, the outer perimeter of the skeletal remains should be marked in such a fashion as to last well in excess of a year, and the markers should not be easily removed.

One technique that has been successfully used to establish a permanent underwater location is simply to take lengths of aluminum or galvanized pipe approximately 0.5 to 1 inch (12-25 mm) in diameter, and using a heavy sledge hammer pound four lengths of pipe into the bottom to mark the four corners of a rectangle that clearly defines the underwater recovery site. These pieces of pipe should be set deep into the bottom so that they may not be moved easily. In shallow areas where foot traffic

is present, such as public swimming beaches, etc., they may be hammered down level to or even below the bottom silt/sand. Once this is accomplished, topside photographs may be taken of the surface buoy for relocation in the future. If detailed photographs and measurements are taken, the relocation of the four corner posts will be a relatively simple procedure using an underwater metal detector or by simply probing the bottom with the hands.

Caution: In the recovery of any skeletal remains it is mandatory that the exact location of the body be marked so that the dive team can return to the site for a further search even years after the initial recovery. It is not uncommon for skeletal remains to be identified and connected with a crime months or years after their initial recovery. Because a murder has no statute of limitations (in most jurisdictions), a search may go on for years. Remember to always have a legal standing to search the area where underwater operations are conducted. Have a search warrant or consent to search if the dive site is not located in a public area. Don't lose evidence due to an illegal search.

On the Scene: Mark Your Location

A group of sport divers who were diving in a remote location accidentally discovered the skeletal remains of what appeared to be a human body, in approximately 30 feet (9 m) of water below a steep cliff. Not knowing what to do, they initially marked the location by spray-painting a white X on the cliff, and then they contacted the local police department.

The local public safety dive team attended the following day and recovered the human remains, which were still held intact by a pair of pants and the remnant of what appeared to be a shirt. The remains were removed using an appropriate mesh-bottom body bag, and the entire sequence was photographed using an underwater camera.

The identity of the body was still in question three weeks later when the photographs were being studied, and in one of the photographs what appeared to be a wallet was seen hanging loosely from the pants of the deceased.

Because the dive team had carefully marked the underwater location, a return to the site with

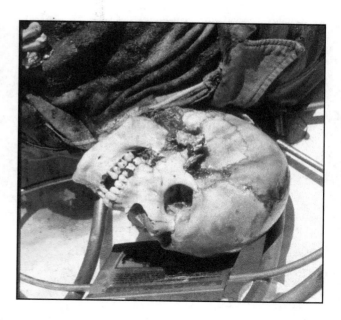

During the recovery of any skeletal remains, the site should be permanently marked and the remains photographed. Great care must be taken during the recovery of skelatal remains. Recovering small remains is not a pleasant task, but certainly it is one that deserves a great deal of caution and dedication by the public safety diver.

bright, surface-supplied lighting was easily carried out. A meticulous, inch-by-inch (cm-by-cm) search of the 5- by 10-foot (1.5 m x 3 m) area failed to locate the wallet, but other artifacts were recovered that assisted in identifying the deceased as a potential suicide victim who had been reported missing 10 months earlier.

In their slow, careful search, the divers had located a pair of eyeglasses that contained the same prescription as that of the missing person. Buried in several inches (~5-10 cm) of silt, however, was the final clue. Several small bones, including the metacarpal bones of the ring finger of the left hand, were located. Still encircling one of these bones was a wedding band bearing the inscription, "Always Together." The deceased, who was identified by these means, had lost his wife to cancer only one month prior to his disappearance. His wife wore a wedding band with an identical inscription.

SKELETAL REMAINS: WHAT DO THEY TELL US?

In recent years there has been a great advancement in the field of forensic anthropology. With this advancement, however, there have been many myths and misconceptions regarding the ability of the forensic anthropologist/pathologist to tell the history of an individual merely through examination of the skeletal remains.

The skeleton is merely one piece in a puzzle that when put together will still reveal only a limited amount of information. While the skeleton is a useful tool for determining facts about the individual, his life and death, it will never equal the whole body as an exhibit that has a tale to tell. Perhaps the value of the skeleton as an investigative lead has been overrated merely because, through modern advancements in this science, it is now possible to glean valuable (and often conclusive) information from an artifact that was worthless in the past.

The human skeleton can relate specific information, such as the following.

Human or Animal?

When only a few bones are found, this is often not an easy question to answer conclusively. This becomes even more confusing when it is understood that the limb bones of some of the smaller animals bear a strong resemblance to those of a child. In addition to this fact, certain animals have skeletal similarities to humans. The foot of a common North American black or brown bear appears identical to the foot of a human, and its true origin is often revealed only under close inspection by a qualified examiner.

When there is doubt, often the most direct route of identification is to compare the found bones with those of known local animals.

The recovery of any skeletal remains entails not only the recovery of all human tissue but also a detailed search and recovery of all artifacts found at the site. In this case, the victim and his clothes had become so badly decomposed that these articles were located after his removal. A painstakingly slow search was conducted in the immediate areas for anything that remained. In cases such as this, the use of underwater metal detectors is advisable. Both keys and a medallion found buried in the silt at the scene contributed to the ultimate identification of this victim.

How Many People?

While it would be a very rare event for a public safety dive team to be involved with a multiple-skeletal recovery, such could easily be the case if the remains were the result of an air crash, fatal motor vehicle accident, or even a multiple drowning in which the bodies were not recovered for a considerable length of time.

Often conclusive proof comes that more than one skeleton has been recovered when more than one bone of the same type is recovered. One large left femur (the long leg bone connecting the knee to the hip) and one small right femur would certainly indicate the presence of two human bodies, in the absence of gross deformity of either bone.

All bones must be recovered — even the small, perhaps insignificant ones. In addition to recovering all bones present at the site, the dive team must be willing and able to collect any and all material that even remotely seems to be foreign to the bottom. What may appear different but not truly resemble a bone or collection of tiny bones could represent a fetus in the early stages of development. Complicated by the fact that these tiny remnants may be found in, around, and confused with larger, more retrievable skeletal remains, it's even more important that everything be recovered.

Determining Sex From Skeletal Remains

Sex can be determined from skeletal remains, but like most findings, the more complete the skeleton presented for examination, the greater the degree of accuracy. The bones most commonly used in determining the sex of the individual are as follows:

(a) The pelvis and its accompanying bones: the sacrum (the large triangular bone situated at the lower part of the vertebral column), ilium (the large basin-shaped bones that form the greater part of the pelvis), the ischium and the pubis (the structural portion of the pelvis forming the lower areas)

(b) The femur (thigh bone). In males, the vertical diameter of the head is about 1.8 inches (45 mm) in males and 1.7 inches (42.4 mm) in females. Generally, a diameter of less than 1.6 inches (41.5 mm) is almost certainly considered to be a female; similarly, a diameter greater than 1.8 inches (45.5 mm) is almost certainly of male origin.

(c) The sternum. The sternum is composed of three distinct bones (or pieces). Situated at the top is the first piece, called the manubrium, followed directly below by the second piece, the gladiolus. Directly below that is the third and smallest piece, the ensiform or xiphoid appendix. In the male, the length of the manubrium is considerably less than in the female.

(d) The skull. In most instances, the lower jaw of the female is noted to be smaller and lighter than that of the male. In addition, in the female there is most often a uniform, graceful curve. The superciliary ridge can also be used to indicate the sex of the skeleton. The superciliary ridge can be felt, low on the forehead, just above the eyebrows. It follows the contour of the eyebrows, arching upward from the base of the nose, and becomes less distinct as it continues to arch outward. This ridge is usually prominent in the male but slight or even apparently absent in the female.

Determining Age from Skeletal Remains

In the average male, the skeleton is not fully developed until approximately age 25. Females, on the other hand, usually exhibit a fully-developed skeletal system at the age of about 23 years.

Up to these ages, the actual age of the victim can usually be fairly accurately determined from the degree of ossification (hardening and complete bone formation) of the skeleton. The bones most commonly used in determining the age of the victim are as follows:

(a) The skull. By studying the closure of the various fissures (fontanelles), the forensic anthropologist/pathologist can determine the age of the victim fairly accurately. These fissures, depending on their location, usually are obliterated (closed by normal bone growth) from the age of three months to eight years. Since these fissures do not all ossify (develop into bone) at the same rate, an intact skull of a child less than eight years old will reveal an approximate age.

(b) The spine (vertebral column). At birth, each vertebra (except the sacrum) are composed of three distinct pieces. The vertebra are fully ossified upon reaching adulthood.

(c) Other bones. Depending on their location, under close examination certain bones will reveal an approximate age of the victim.

Caution: It is very difficult for a forensic pathologist to determine the age of an individual when given only limited skeletal remains for examination. As much of the skeleton as possible must be recovered and presented for examination, since various elements will only confirm an approximate age.

For example, the bones of the fingers will yield accurate age estimates for very young children, while pelvis, leg, and arm bones are more accurately used to determine the age of a teenager. When using skeletal remains to determine the age of a victim, the examiner will use various bones and sites on these bones to substantiate his findings. Like any evidence, the more of it there is, the more definite the findings will be.

Determining Height from Skeletal Remains

If the skeleton is recovered in a complete or nearly complete form, the approximate height of the individual can be calculated by reassembling the structure using a vertebral column support. Intravertebral discs are often fabricated out of plasticine or modeling clay. The vertebrae are then aligned and pushed together until the spinal processes are properly aligned.

When only limb bones are recovered, an approximate height can be calculated using special tables.

Determining Ethnic/Racial Origin of Skeletal Remains

Because of the amount of racial mixing present in modern man, the determination of an exact racial or ethnic background merely from the examination of skeletal remains is at best an inexact science. There exists published literature on this science, but contrary to popular opinion, the determination of the exact racial origin of an individual from the examination of the skeletal remains is not an easy task; indeed, it is usually impossible.

Determining the Postmortem Interval from Skeletal Remains

Due to the large number of variable factors present, the examination of skeletal remains gives very little information as to the time of death, or the postmortem interval. Factors such as the degree of moisture present, the bacteria present in the medium in which the body was found, the presence of insects and animals, etc., all affect the skeleton, making the determination of a postmortem interval nearly impossible.

Determining Physical Trauma from Skeletal Remains

Since the skeleton is a hard (often brittle) structure, it will record for later examination many types of trauma. When an intact skeleton is recovered, blunt trauma from deaths such as a motor vehicle accident, beating, fall, etc., may be recorded on the skeleton. Specific types of fractures are indicative of specific types of injuries, and a pathologist is often able to examine a specific type of fracture and relate its origin. For example, a body that has been badly burned in a fire and then disposed of in water may display a specific type of cranial (skull) fracture that would indicate to the lay person a blunt trauma injury but would serve as proof to the trained pathologist that the victim was exposed to a very high-heat environment. Thus, the skull of a victim found underwater could convey to the investigator that a fire was involved, either as a cause of death or as a failed attempt to dispose of the body prior to submergence in water.

USE OF RADIOGRAPHY (X-RAY) IN SKELETAL EXAMINATION

All skeletal remains should be X-rayed prior to disassembly. A complete X-ray examination may provide many clues, such as the following.

(a) The number, location and shape of dental fillings. In the examination of a child's skeleton, the development of the various teeth will yield very accurate information as to the age of the victim.

(b) The age of the individual by overall ossification of the skeleton can be estimated.

(c) The presence of various fractures or medical procedures may be revealed by X-ray examination. These may be compared to the known medical history of a suspected victim for verification of his identity.

(d) Specific identification. Since the bone structure of all individuals is different, if previous medical X-rays are available on an individual who is suspected of being the victim, a comparative study may be made.

In cases where the X-rays of the recovered skeletal remains are to be compared to known X-rays taken during life, care must be taken to take the postmortem X-rays in exactly the same position and at the same magnification as those take prior to death.

The Teeth as an Indication of Age

It is very important that all teeth present in and around the skull be collected and presented to the medical examiner. The following chart indicates the age of an individual as various teeth erupt. This is a determinant of age.

Ages at which Teeth Erupt

Deciduous (Baby) Teeth

Central Incisors	6 months
Lateral Incisors	8 months
Canines	17 months
First Molars	13 months
Second Molars	22 months

Permanent (Adult) Teeth

Central Incisors	7 years
Lateral Incisors	8 years
Canines	9 years
First Premolar	10 years
Second Premolar	11 years
First Molars	6 years
Second Molars	12 years
Third Molars	20–24 years

Conclusion

The recovery of skeletal remains demands a meticulous approach by the public safety dive team. Everything must recovered, because even the smallest of bones may be able to reveal information that other artifacts cannot. Indeed, even the smallest bones of the body, found within the middle ear (malleus, incus, and stapes, commonly referred to as the hammer, anvil, and stirrup) may reveal information as to the age and even lifestyle of the victim. All artifacts must be recovered.

After removing skeletal remains, dive teams should consider the benefits of sweeping the area with a metal detector before leaving the site.

6 BODY RECOVERY REPORTING PROCEDURES

Body recovery and the resulting death investigation is composed of many steps: finding the body of the accident/homicide victim, followed by observing your find and analyzing the postmortem changes. The next logical step in this procedure will be preparing a clear, concise, court-ready report.

WHY PREPARE A COURT-READY REPORT?

The preparation and submission of a court-ready investigative report not only forces the public safety dive team to investigate and record its dive operation/mission in a logical, clear, and concise manner, but also it will serve as a future reference for either criminal or civil court proceedings. The preparation and submission of a court-ready report will ensure acceptance by the sponsoring police department/agency and will be instrumental in building the reputation of your dive team as a professional investigative unit, raising its credibility above and beyond that of a mere recovery team.

THE CONTENTS OF YOUR REPORT

A good report should contain, in detail, the four Ws: who, what, where, and when. The why, of course, is answered by the courts. In the event of a criminal activity, the criminal court having jurisdiction will convene a hearing or trial to determine the answer to these questions. In the event of an accidental death, either the coroner's court or a civil court of inquiry will convene to elicit and record details and ultimately record the incident in a legal format.

In brief, the report submitted by the dive team should contain everything that the dive team saw and did in connection with the dive mission. While opinions may have a value, if they are submitted objectively and without bias, they should clearly be stated as opinions and attached to the report

separately. Opinions are not always solicited by the courts; indeed, often they are not even allowed into the court record unless certain conditions have been met first. Primarily your report should read as your evidence. It should contain only what you actually saw and did. The information the dive team was given prior to the operation can form an integral part of your dive log, which is a report separate and distinct from your police report.

The following pages will offer a sample or skeleton forensic report to follow. It may be photocopied and utilized in the format presented or modified to suit local needs.

Since this report may be subject to the scrutiny of the court, its accuracy will be of primary importance. For this reason, it is recommended that this report be completed in handwriting during the mission or actual diving operation. This handwritten report can then serve as your notes made at the scene. In the submission of verbal evidence during court proceedings, it is usually permissible to refer to notes only if they were made at the time of observations or very shortly afterward. Your report to your sponsoring police department can be made at a later date, using your field notes as a guide.

THE BODY RECOVERY REPORT

All public safety dive teams should develop clear, concise and accurate reporting procedures. In many cases, the form for the report is supplied by the requesting agency. In other cases, dive teams develop their own specific reporting forms based on local requirements. The actual format of any report is not the critical issue — indeed, there are almost as many reporting forms as there are agencies. What is of primary importance is that all critical information be recorded. Information in many cases becomes evidence. To preserve information becomes one of the primary responsibilities of the investigative dive team — to preserve evidence.

The following first page, along with the subsequent checklist, Supplemental Underwater Recovery Report — Body Recovery, Postmortem Observations, is a suggested guide. This checklist will not only enable the dive team to report clearly and objectively, but also provide a reminder of the critical observations and records to be made at the scene of any body recovery.

The purpose of this checklist would indeed be fulfilled if it were photocopied and taken into the field as an integral part of the investigative dive team's basic equipment. This is not meant to replace existing reporting formats, merely to supplement them.

UNDERWATER BODY RECOVERY REPORT

Submitted by _____

_____

Address _____

_____

_____

Phone number _____

_____

Services requested by (name of police officer, etc.) _____

_____

_____

Department: _____

Date and time of request: _____

Case heading and file # _____

Dive team members on scene: _____

_____

_____

When evidence/body was recovered: _____

_____

_____

SUPPLEMENTAL UNDERWATER RECOVERY REPORT
Body Recovery Postmortem Observations

1. Name of deceased _____

2. Antemortem health _____

3. Recovered in ❏ Fresh water ❏ Salt water

4. Water type Clean 1 2 3 4 5 6 7 Polluted

5. Visibility (feet/meters) _____

6. Depth (feet/meters) _____

7. Water temperature at depth _____ °C _____ °F

8. Ambient temperature _____ °C _____ °F

8. Body position ❏ On bottom ❏ On surface (floating)
    ❏ Face down (prone) ❏ Face up (supine)
    ❏ Left side down ❏ Right side down
    ❏ Other: _____

9. Degree of bloating Nil 1 2 3 4 5 6 7 Severe

10. Tissue decomposition Nil 1 2 3 4 5 6 7 Severe

11. Presence of animal feeding (anthropophagy) Nil 1 2 3 4 5 6 7 Severe

12. Discoloration of tissues (livor mortis) degree and location _____

_____

13. Contact lividity (white spots due to pressure on the skin) _____

    Location _____

14. Cadaveric spasm (death grip) ❑ YES   ❑ NO

15. Rigor Morti: ❑ YES   ❑ NO
    Location   Neck   Nil 1 2 3 4 5 6 7   Severe
                Elbow   Nil 1 2 3 4 5 6 7   Severe
                Knee   Nil 1 2 3 4 5 6 7   Severe

16. Cutis anserina (goose bumps): ❑ YES   ❑ NO

17. Cyanosis   ❑ YES   ❑ NO
    Lips   Nil 1 2 3 4 5 6 7 Severe
    Elsewhere   Nil 1 2 3 4 5 6 7 Severe
        Specify _____

18. Ocular (eye) changes _____

        Cornea   Clear 1 2 3 4 5 6 7 Opaque

19. Travel abrasions ❑ YES   ❑ NO
    Location _____

20. Wauschaut (dishpan hands effect)   ❑ YES   ❑ NO
    Location _____

21. Presence of foam   Mouth   ❑ YES   ❑ NO
                        Nose   ❑ YES   ❑ NO
                        Blood present ❑ YES   ❑ NO

22.	Suspected time in water (drowning to recovery) _____

_____

_____

_____

_____

23.	Water samples taken	❑ YES	❑ NO

Where (location and depths) _____

_____

_____

_____

24.	Trauma/wounds (describe and give location to complement attached diagrams) _____

_____

_____

_____

_____

25.	Remarks (include description of clothing, jewelry, and unusual features) _____

_____

_____

_____

_____

_____

_____

_____

When recording observations of physical trauma, diagrams are essential. The following schematics will assist in explaining the location, type, and severity of trauma to the deceased's body. The schematic should be taken to the scene, and observations made at the scene should be transcribed (drawings, including notes) directly onto the body outlines. These will serve as an accurate reference and notes made at the scene for future reference.

All notes and comments may be handwritten directly onto the page displaying the specific trauma.

PHYSICAL TRAUMA NOTES

Case # _____   Name: _____

Prepared by: _____

Date: _____   Time: _____

PHYSICAL TRAUMA NOTES

Case # _____   Name: _____

Prepared by: _____

Date: _____   Time: _____

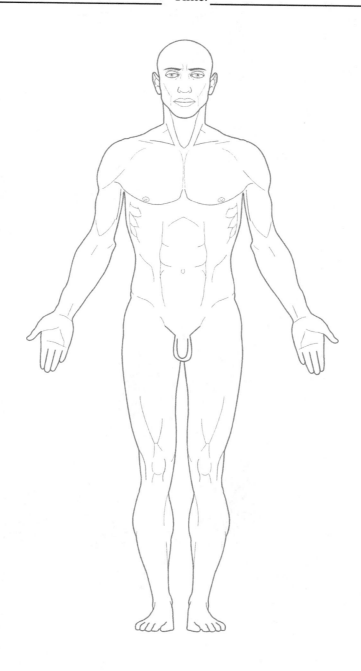

PHYSICAL TRAUMA NOTES

Case # _____ Name: _____

Prepared by: _____

Date: _____ Time: _____

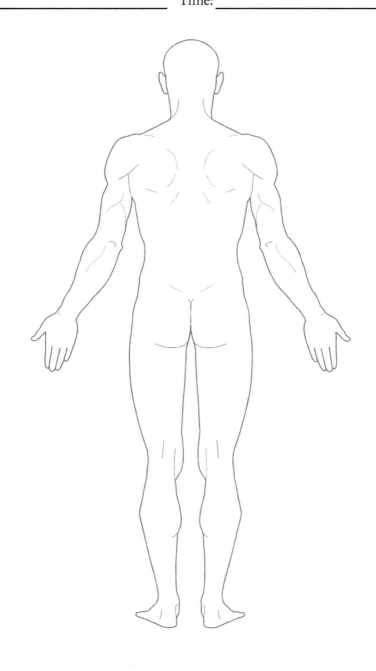

PHYSICAL TRAUMA NOTES

Case # _____   Name: _____

Prepared by: _____

Date: _____   Time: _____

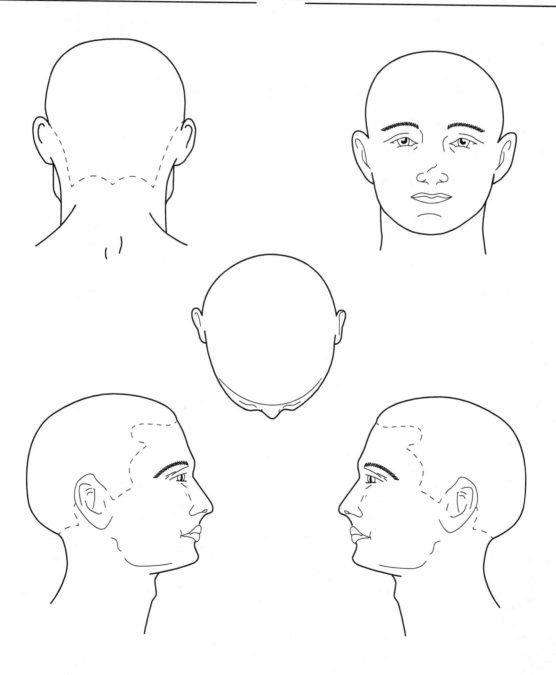

PHYSICAL TRAUMA NOTES

Case # _____   Name: _____

Prepared by: _____

Date: _____   Time: _____

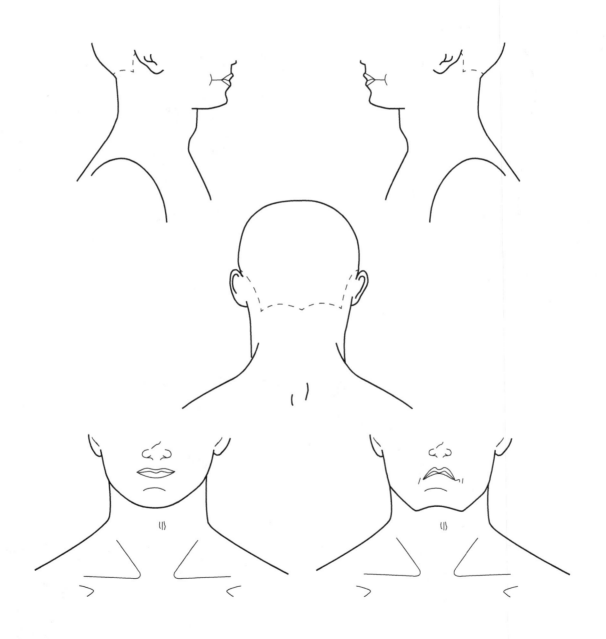

PHYSICAL TRAUMA NOTES

Case # _____   Name: _____

Prepared by: _____

Date: _____   Time: _____

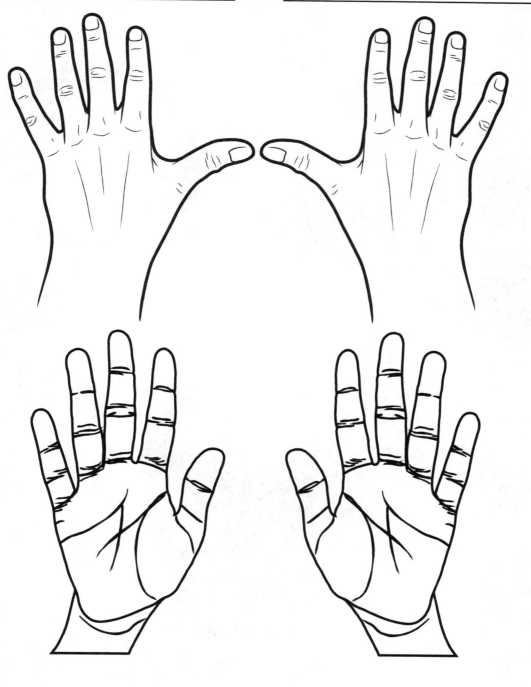

PHYSICAL TRAUMA NOTES

Case # _____   Name: _____

Prepared by: _____

Date: _____   Time: _____

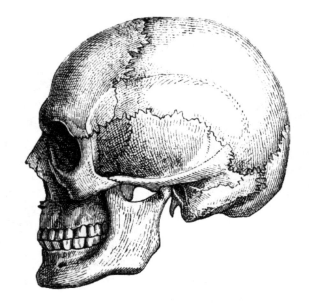
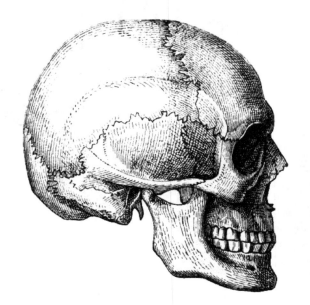
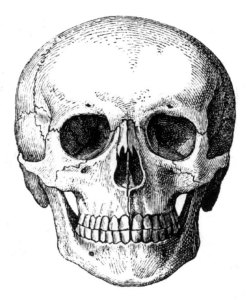

PHYSICAL TRAUMA NOTES

Case # _____   Name: _____

Prepared by: _____

Date: _____   Time: _____

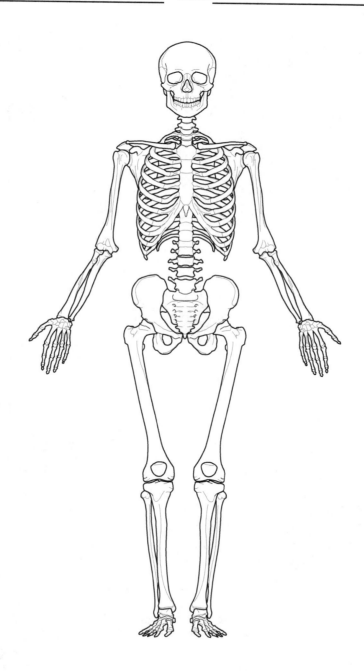

PHYSICAL TRAUMA NOTES

Case # _____   Name: _____

Prepared by: _____

Date: _____   Time: _____

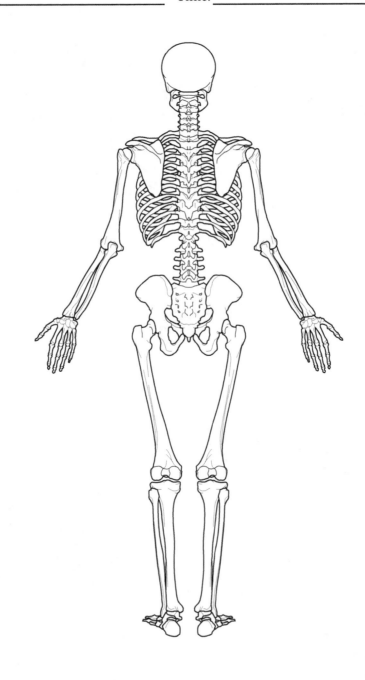

Supplemental Checklist — Skeletal Remains

The recovery of and subsequent reporting procedures involving skeletal remains demands additional detail not normally found in standard reporting formats. The following is suggested as a field guide:

When completing the diagram describing the position or orientation of the various bones located, it may be advisable to cut and paste the various bones or skeletal fragments from the previous pages to prepare a rough sketch of the position in which each bone was found. When the actual bone cannot be clearly identified, the label "unknown bone" is suggested.

As in all evidence recovery, photography is advantageous and recommended; however, even photography cannot and should not be used in place of detailed field notes and diagrams.

SUPPLEMENTAL BODY RECOVERY CHECKLIST
Field Recovery Form: Skeletal Remains

1. Identity (circle):   unknown   known   suspected

   Sex: _____   Race: _____

   Name: _____   Date of birth: _____

2. Describe location of remains: _____

3. Date and time remains recovered: _____

4. Names of all present at recovery site: _____

5. Description of nonskeletal remains (e.g., clothing, jewelry, personal items, etc.) _____

   _____

   _____

   _____

   _____

6. Observations of (suspected) skeletal trauma (e.g., broken bones, suspected cremation attempt, etc.)

    _____

    _____

    _____

    _____

7. Orientation of limbs/skeleton

    To properly document the orientation/position of the skeletal remains, a cut-and-paste diagram composed of elements from the skeletal structure diagram on a separate page is suggested.

8. Other observations: _____

    _____

    _____

    _____

    _____

    _____

    _____

    _____

7 SCUBA FATALITY — ACCIDENT INVESTIGATION

OVERVIEW

The investigation of a scuba diving fatality presents its own unique problems to the investigator. When these problems are understood and the investigator possesses a good general knowledge of scuba diving, the pieces of the puzzle often can be put together in a meaningful manner. Clearly, the investigation must begin with the interviewing of the surviving diving partner(s), followed by the equipment examination, body recovery, and autopsy results. If any one of these phases is missed or poorly completed, the results of the entire investigation should be held suspect and subject to scrutiny. No conclusions can be made without all facts being clearly represented and combined in an accurate, clear and logical manner.

After the investigation has been completed, a report documenting the entire investigation should serve as a clear reference with which even the nondiver may review and understand the incident.

Role of the Investigator

The role of the investigator in a scuba fatality is clear: to investigate. During the investigational procedure, beginning with witness interview and culminating with the final report, resist all temptations to place blame. The attachment of blame in any death (accidental or homicidal) investigation is clearly the responsibility of the courts. After the completion of the investigation, the report that is submitted will be scrutinized by criminal courts, civil courts, coroner's courts and civilian accident review boards who will make their own decisions as to blame or culpability.

The investigator's responsibility is to explain the accident. If the explanation should include personal or professional opinion to fill in missing detail, such opinion should clearly be expressed at the conclusion of the report, and this opinion should clearly be identified as such. Opinion may carry a great deal of weight, but it is never fact.

Scuba Fatality Investigation — The Procedure

From start to finish, the investigation is composed of steps or stages. Although these steps are connected and will combine to form an ultimate outcome or conclusion, they will be discussed separately. They are:

1. The Witness Interview

2. Locating the Body

3. Equipment Examination

4. The Autopsy

5. Formulating a Conclusion: A Reason for the Fatality

6. The Scuba Fatality Report

I. THE WITNESS INTERVIEW

As with most body-recovery missions, the witness interview is not only the first but also the most important part of the investigation. Proper witness interview will supply the dive team with the information necessary not only to locate the body but also to understand what has happened and how the accident actually occurred. These details, which must be solicited during the witness interview portion of the investigation, may show certain discrepancies. When the whole picture is examined, the discrepancies may reveal an even more accurate picture of the accident than the witness was able (or willing) to relate.

Fortunately, the scuba diving fatality accident nearly always supplies the investigative team with a witness. With the common acceptance of the buddy system in sport diving, rarely is there a missing diver reported with no last-seen point or last known activity to afford the dive team with the bare essentials necessary for an effective search.

Unlike most other witnessed drownings, however, the survivor of a scuba fatality is a witness with a difference: In most cases, there will be a feeling of remorse and guilt encountered when interviewing the survivor. The underlying feeling of guilt of having let his diving partner die in most cases leaves the survivor with a deep feeling of inadequacy. During their training, divers are taught that buddy diving prevents accidents; when an accident occurs, it is felt that one or both of the divers have failed in the role of buddy. In short, by the very fact that there has been an accident, the survivor will in all likelihood feel as though he has failed.

In addition to this feeling of guilt and remorse, he may even suspect that he will be held accountable in a coroner's, criminal, or civil court of inquiry. This underlying feeling of guilt may necessitate a special type of interviewing skills, since even if the person who survived the accident is not suspecting criminal or civil responsibility, the emotional burden of being the only survivor may cause him to subconsciously change or color the story to one that places more blame on the deceased.

The purpose of the interview is not to attach blame — it is to collect information that will enable the dive team to locate the body. Blame (if it is to be found) will be as a result of a further investigation and ultimately the attachment of blame is the responsibility of the judicial system. The role of the dive team is solely an investigative one.

Bearing this in mind, the following points are suggested for an effective and detailed interview.

Privacy: The interview should be conducted by one (certainly no more than two) dive team members. This will help the survivor avoid feeling that he is being overpowered or "ganged up on" by the investigators. Ideally, the interview should be conducted in private by only one dive team member. This will help to establish a feeling of confidentiality and trust wherein the survivor may find it easier to relate intimate details of the dive and even offer suggestions or feelings as to how, where, and why the accident occurred. When there is only one dive team member conducting the interview, the survivor may feel less like he is being interrogated and more like he is assisting the dive team in finding the body of the

lost diver quickly. In truth, that is the primary reason for the on-site interview.

Privacy may be afforded in a nearby vehicle or any convenient location where there is neutral territory. Both the dive team truck or van and the survivor's vehicle should be avoided, since neither are neutral. The exceptions to this rule would be if the survivor is a strong, controlling type of personality; then the authority and credibility of the investigative dive team may be established (indirectly) by conducting the interview in the department's vehicle.

Conversely, where the survivor is badly shaken by the incident and is obviously feeling frightened or deeply remorseful, his own vehicle may be a desirable location for the interview. Whatever the physical location of the interview, it should be one that places the interviewer and the survivor in a private, almost intimate setting. The location should be close enough to allow for easy communication with the on-site dive team but far enough away so that the survivor is protected from the news media and casual conversation of the attending police, fire department, or members of the general public, which may be overheard. One misdirected comment accusing the survivor of blame in the accident could easily terminate what could otherwise be a meaningful and fruitful interview.

The interviewer: The dive team member selected as the interviewer should ideally have some experience investigating scuba fatalities. However, since this is not often possible, it would at least be advisable that the dive team member chosen be an individual who is not only a very experienced diver but also an individual who possesses empathic, understanding qualities — an individual who is able to put people at ease and conduct an interview in a friendly, warm and understanding manner. Above all, the interviewer should possess the understanding that what has just occurred may truly be an accident and that the survivor is a secondary victim of the accident, not a suspect.

Length of the interview: Most dive teams are eager to enter the water and commence their search. Experience has shown conclusively that thorough preparation not only shortens actual search time but also increases the success rate of all searches. For this reason, the length of the interview is not fixed. The interview should be as long as it takes to obtain enough details to allow the dive team to effectively and safely search for the missing diver.

One addition to the interview not commonly thought about is the care of the survivor after the initial interview. This is worthy of consideration, and will be discussed under the heading "Concluding the Interview."

Details to be Obtained: Most interviews seek to reveal the 5 Ws: who, what, where, when, and why. The initial interview of the survivor of a diving accident will concentrate on all but the why.

Who: This portion of the interview will help to ascertain not only the identity, name, age, and address of the victim but also diving experience and previous diving history if it is known.

What: In determining what happened during the fatal moments of the dive, the scenario may be reconstructed to determine if the victim will likely be located at or near the last-seen point. Often there was a mere loss of buddy contact, and the victim was not noticed as missing for several minutes. This fact alone may broaden the primary search area. In determining what happened, the last known activity is of primary importance. Was there a rescue attempt? Was there an attempt at sharing air? Was underwater entanglement involved? Was the lost diver complaining of any problems or discomfort prior to the dive? What activity was he involved with? Was he carrying anything? These are but a few of the possible questions, which are as varied as the number of scuba accidents that occur.

During the "what (happened)" phase of the interview, the attachment (or even mention of) blame should be strictly avoided. The ground rules of the interview may have to be set early if resistance is met so that the survivor realizes that his sole role is to assist in locating the lost diver.

Where: The "where" portion of the interview should reveal whether the accident occurred on the bottom, during ascent or descent, or on the surface. The primary importance of this portion of the interview is in determining a last-seen point. As investigators, it is best to have witnesses show you at the physical location rather than simply tell you where they think the incident occurred. Often locations appear differently when physically seen in person by the witness. If the last-seen point was on the bottom, specific details or landmarks should be solicited. To assist the recovery team in locating these landmarks, other bottom topographical structures or landmarks that were seen prior to those at the last-seen point should also be queried. For example: If the last-seen point was a large rock outcropping, what was seen on the dive prior to this? This may assist the dive team in laying a track or following the course of the actual dive in order to follow the path of the divers prior to the actual incident.

If the survivor surfaced shortly after losing contact with his partner (or if contact was lost on ascent), the last-seen point should be provided via a surface visual reference. To determine this last-seen point, it is advisable to take the survivor out (preferably in a small craft) and have him lead the interviewer to the exact point. At that time, the survivor may be asked to place a marker buoy at the point where he surfaced. A small craft is the best choice for this relocation of the last-seen point, since it is less traumatic than placing the survivor in a situation where he may be asked to swim over the body of his lost friend; also, a small craft will place him close to the water's surface and make his visual reference to the shoreline closer to what he experienced when he was swimming than if he were placed in a larger boat.

The survivor should be encouraged to give the directions and drop the marker buoy where he remembered surfacing; the interviewer should avoid all temptation to guide, suggest, or direct the dropping of this marker buoy. The simple direction to "drop the buoy at the exact location where you surfaced" will give the best results. The survivor should only be reminded that he is to drop this marker where he surfaced and not where he thinks the lost diver may be. If the latter is suggested by the survivor, a secondary marker may be used for the suspected location of the lost diver. The buoys should be marked differently to avoid confusion during the search. The practice of having the survivor place a marker buoy has been found to be one of the most productive and important procedures in the chain of events that leads up to the successful recovery of the missing diver.

When: Perhaps one of the first questions should be, "When?" This will determine whether a recovery or a rescue operation is in effect. Once it has been determined that a rescue is no longer feasible and that the mission has become a recovery, the "when" may help to determine the amount of air remaining in the lost diver's cylinder. This may indicate the distance traveled after buddy contact was lost. When inquiring into the "when" portion of the interview, more important than the actual time or duration of the dive is the amount of air remaining in the survivor's cylinder when loss of buddy contact was first noticed. The interviewer should also attempt to determine whether the lost diver typically consumed air more rapidly or more conservatively than the survivor. If from these details it can be suspected that an out-of-air scenario precipitated the accident, then the last-seen point will be of crucial importance.

In determining the "when" of the accident, the pieces of the puzzle may begin to be placed together. If the "when" was in the first few minutes of the dive, the body will likely be located very close to the point of entry into the water. Conversely, if the "when" occurred well into the dive and considerable swimming activity was involved, a large search area may be involved.

In short, the answer to the question "when?" is merely an attempt to retrace the steps of the dive in an effort to locate the lost diver somewhere along those steps.

Concluding the interview: When the interview is to be concluded, the interviewer should ensure that the name, full address, and telephone number of the survivor is obtained and recorded. It should also be determined whether or not the survivor is on vacation or permanently resides at this location, since it will likely be necessary to contact him at a future date. The interview should always be concluded with a positive relationship having been established. The survivor should always be made aware of the fact that no matter what has happened, his assistance was invaluable and that he has helped. This relationship between the interviewer and the survivor will ensure that should any contact be necessary in the future (and it often is), there will be no barriers to

communication. While it may seem that future care of the survivor is not the responsibility of the dive team, many fully functioning dive teams will consider the survivor as a secondary victim and assist him with details such as transportation home, emotional reassurance that he did all he could, etc. Any future care or contact with the survivor, of course, should be considered only after consulting with the police department primarily responsible for the investigation of the fatality.

Making notes on the interview: During the preliminary phase of the interview, the display of a notebook and pen may be counterproductive. A good personal rapport is essential, with the interviewer developing a relationship of trust and concern. Often the interview will progress quickly to a recollection by the survivor of just what has happened. After the story has been related and the interviewer is satisfied that a good rapport has developed, it is then advisable to make notes of the details of the conversation.

Prior to producing a notebook, it should be explained that the interviewer's intentions are to document details so that he can accurately relate these details to the dive team. It is also advisable to seek the survivor's permission to make notes; although this is not necessary in a legal sense, it at least shows the survivor that you care enough to ask. This mere asking for permission will also assist in solidifying a position of trust and caring between the interviewer and survivor.

When making notes, take care to record the date, time, and location as well as the names of all people present during the actual interview. The production of these notes will likely be required at a later date for court hearings or the investigational report. The notes should be retained in the possession of the interviewer, but copies (if requested) may be made available to the investigating police department. In the event of future court proceedings, notes made at the scene will invariably be summoned along with the interviewer.

During the actual process of taking notes, it is always advisable to record actual phrases used by the survivor so that the story may be told "in his own words," but the entire conversation is rarely recorded in a handwritten format unless a formal statement is being taken. It should be remembered that the interviewer's primary responsibility is to collect facts that will help to understand the accident and locate the body of the victim.

The use of tape recorders during an interview is often not advisable, though in some regions of the United States an audio and video recording of an interview where a homicide is contemplated is required and even mandated by some departments. In many countries, the use of a concealed audio recorder is against the law, but in areas where it is legal to utilize a hidden audio/video recording device, ultimate disclosure may violate the trust placed in the interviewer. When permission is sought to record the conversation, in most cases, the survivor will feel quite inhibited, and unless excellent interview skills are present, the focus of attention (by the survivor) will be on the recording device. All attention should be focused on the development of rapport between the survivor and the interviewer and an accurate recollection of what happened during the accident. Although they are often used, an audio recorder can never replace good interviewing skills.

2. LOCATING THE BODY

Certainly more than any other activity, indeed even more important than complex search techniques, the witness interview is the most important aspect of the investigation. Without an adequate interview conducted with the survivor by an experienced dive team member, chances of success are at best poor.

In determining a search pattern, the question "Where do we start?" must be answered by determining a last-seen point. Of course, this may not lead directly to the location of the missing diver, since divers, unlike other drowning victims, can breathe and travel underwater. Still, the adherence to a strict policy of beginning at the last-seen point will invariably increase the success rate for this type of diving operation.

In most cases, the fatality seems to follow quickly after the loss of buddy contact, even when the loss of such contact was seemingly casual or unimportant at the time. This may be partly due to the fact that many survivors find it difficult to remember that they lost contact during a time of need, or it may be coincidental. Because of this, however, the use

of a marker buoy (set in place by the survivor) may serve as a focal point of an expanding circular search using lines or a grid search using boat-towed underwater sleds. As in all searching, free-swimming search teams rarely yield good results unless the underwater visibility is excellent or the search area is small.

One possibility that should not be overlooked, especially when an accident has occurred in an area where strong currents are present, is the possibility of a distressed diver on the surface at a considerable distance from the accident site. Although not common, there have been many instances where a diver has been located by airplane or search boats several miles/kilometers away from the initial dive site. As in all searches, consider all possibilities.

When searching for a lost diver where excessive depth is involved, there are alternative search techniques that may be used. These search tools may provide the dive team with a degree of success not commonly found if they are used carefully and by trained personnel.

ROVs: Remotely operated vehicles (ROVs) are an asset when the search area is deep, visibility is acceptable, and the search area is reasonably contained. The use of ROVs is expensive, and success is more dependent on the experience and ability of the operator than on the capabilities of the vehicle itself. An ROV is nothing more than a video camera with its own propulsion system, tethered to the surface by a cord that conducts power to the propellers and a signal to the topside monitoring video camera. It must contain some device for retrieval of the lost diver, even if that device is a simple makeshift grapple hook placed out in front, within the viewing area of the camera.

The operator must not only be capable of deft manipulation of the vehicle (to avoid stirring up a silt bottom) but must also understand and be able to operate his ROV within the confines of a good search pattern. In addition to a topside video monitor, a real-time videotape of the entire search should be made for future viewing, since it may provide details not previously known or seen either by the survivor or during the search.

Magnetometers: Boat-towed magnetometers or metal detectors may assist in searching a large area if the diver was wearing a steel cylinder and the bottom is relatively clear of metallic (iron) debris.

The use of metal detectors is often a great asset when searching for a lost diver under ice. In ice diving fatalities, the victim is usually located up against the ice canopy. Search patterns conducted by personnel walking on the ice using metal detectors will often meet with success and reduce the amount of time divers must spend in this hazardous environment. Certainly the most effective tool for searching for a diver lost under ice is a rotating-type sonar that can be lowered through a hole that is augured through the ice.

Magnets: The use of large magnets to locate a diving fatality victim in deep water is, at best, better than nothing. Most often the only piece of equipment capable of being attracted to such a magnet is the steel diving cylinder. This type of tool is not often successful, since the cylinder is round and affords a very poor contact for a magnet, and most diving victims are found on the bottom in a supine (face-up) position. In addition to this, a magnet is a contact instrument that must either come in actual contact with the cylinder or at least come within a few inches (centimeters) in order to be attracted sufficiently to hold.

Sonar: Since the 1990s sonar has been acknowledged as a valuable asset in the underwater recovery field. With the advent of new technology and proper sonar schools, the use of sonar is widely used throughout the underwater recovery community. A lost diver with a metal tank and associated dive equipment can present as an excellent target for sonar.

Sonar allows large areas to be searched using efficient search patterns guided by global positioning systems (GPS) or surface marker buoys. When targets are registered, markers are set or GPS positions recorded, and dive teams are later dispatched to investigate each suspected target. The use of sonar decreases the duration of dive time for the team and, in doing so, increases the overall safety of the operation. In addition, effective use of sonar allows large areas to be searched with incredible accuracy.

Sonar is most suitable for searching bodies of water with relatively flat bottoms; however, as with any search tool, its value and efficiency is determined chiefly by the expertise of the individual operating the equipment. For this reason, if sonar is used in the search for a body or submerged vehicle (car, boat, or aircraft), proper training in its specific use should be considered not optional, but mandatory.

3. EQUIPMENT EXAMINATION

Rarely in the investigation of a scuba fatality is equipment failure the cause of the accident. What may be encountered on close examination, however, is poorly maintained equipment. Poorly maintained diving equipment, if nothing else, is indicative of an inadequate attitude toward safety by the accident victim. Where equipment failure is suspected, a careful inspection is a mandatory phase of the investigation, and this inspection must be conducted by a qualified and neutral person or agency.

In many respects, diving is similar to flying. When an accident has occurred, equipment failure is always suspected but rarely confirmed. As in most flying accidents, pilot error is cited as the chief cause of the accident; similarly, in most diving accidents diver error is ultimately to blame. Even so, adequate equipment inspection is not an option but rather mandatory. What we have come to call diver error may in fact be partly the result of poorly maintained equipment, which was one of many small factors leading to the ultimately fatal chain of events.

When equipment failure is suspected as a contributing factor toward the accident, it is usually in one of a few specific areas. Obvious but worth repeating is the fact that this type of equipment failure is in truth diver error in fitting, use, or maintenance of his diving equipment. The following equipment is often noted to have been a contributing factor in the ultimate fatality.

Mask: When a victim of a scuba fatality is located, there will often be a quantity of blood, froth, and water in the face mask. This should not necessarily be taken as a sign of physical trauma or a poorly sealed mask. Where the fit of a mask is in question, the answer will often come from the history of its use. If the victim was using his own personal mask and did not have a history of complaining about the seal, the integrity of the mask can usually be assumed to be satisfactory. If the mask was loaned, rented, or for some other reason being used for the first time, a poor fit and subsequent problems with acquiring a good seal may be suspected as being a contributing factor. While water in the face mask is seldom a problem for the experienced and seasoned diver, it is very unsettling for the novice and often has quickly led to the termination of an otherwise uneventful dive. The history of the mask should be questioned.

Regulator: Most diving regulators that are on the market, while varying in their ability to perform (especially at depths in excess of 100 feet/30 m), are safe in that they are able to perform according to manufacturer's specifications. While some regulators offer less inhalation and exhalation breathing resistance than may be desirable, few regulators, if any, can be considered unsafe.

The safety or operational capability depends on two main functions, which are sometimes impaired through poor care or servicing. These two functions are (1) the ability of the regulator to maintain a dry first stage and (2) the ability of the regulator to provide easy and adequate air flow during inhalation.

1) Dry first stage: A regulator that is constantly "sipping water" into the first stage can easily be diagnosed by removing the external rubber or plastic exhaust T. This exhaust T, which normally helps to guide the exhalation bubbles away from the diver's face, serves mainly to protect the thin rubber or silicone one-way valve. This valve should be inspected immediately upon recovery of the equipment from the water. A deteriorated exhaust valve is easily recognized by its stiff and nonflexible appearance. It may be visibly pebbled around its outer edge or merely hard to the touch. In addition, it may be seen to have a curl to its normally flat disc.

The easiest way to determine the water-tight integrity of the first stage of a regulator is simply to test it physically. Prior to removing the external rubber exhaust port, the first stage may be lowered into as little as 6 inches (15 cm) of water and, with

no high-pressure air being allowed into the first stage, an attempt at breathing through it can be made. This inhalation should first be made gently then forcibly. This is a common test, usually performed immediately after the recovery of the equipment and prior to further handling or shipping. On one occasion, a small piece of shell was noted to be lodged under the exhaust valve, which caused a noticeable leak into the first stage with every inhalation. This was confirmed with the removal of the outer cover but may have been dislodged during shipping to the testing facility.

NOTE: This on-the-scene test is not a diving procedure. If it is to be carried out, it should be done in a basin of water or in knee- or waist-deep water. Diving equipment that is suspect should never be test dived, except under the most stringent guidelines.

2) Breathing resistance: When unacceptable breathing resistance is encountered, it is rarely a design flaw in the regulator. Most commonly it is a result of a poorly maintained regulator. In the first stage, air that leaves the diving cylinder must first pass through a metallic screen (usually referred to as a sintered filter) before it enters the body of the regulator. The purpose of this filter is to restrict the entry of particles (rust from internal tank corrosion or charcoal from compressor filters) into the regulator. Since this filter is an effective screen, it may slowly become clogged, thus restricting the air flow. This may not be noticeable by a diver who is very experienced and/or relaxed during most dives, but it may become an incredibly strong contributing factor in deep dives or dives in which physical exertion underwater becomes quickly necessary. Since most diving fatalities are preceded by some degree of panic or struggle, this clogged filter may be the final link in a series of events that results in the fatality. In short, a clogged or partially clogged first-stage filter may not be noticed by the diver until the very crucial time where optimum air-flow is necessary for survival.

It is indeed unfortunate that all too often during the equipment postmortem following a diving fatality, the optimum operating capabilities of a regulator come under scrutiny, when in fact the

Human error, not equipment failure, is the cause of a scuba diving fatality when equipment inspection reveals a regulator corroded to this extent. Tests performed on this regulator revealed that its functioning capability was approximately 20 percent of its design parameter. A simple cleaning and servicing, however, restored its function to a point where it exceeded the manufacturer's specifications.

cause of failure (which is often never detected) may lie in the portion of the regulator that is the simplest to maintain and test.

The typical clogged first-stage filter is easily recognized by its dirty, often discolored appearance. It should appear shiny, metallic, and sparkling under bright light.

Dive computer: Any dive computer(s) worn by the victim may provide the most accurate information of what may have transpired during the dive. Most dive computers record basic dive parameters such as depth and time. Such profile information may answer questions such as "What was the last activity of the diver (ascending/descending)?" or "When did the incident occur?" They are often much more reliable and objective than buddy or witness statements.

Other computers may record other valuable information, such as decompression status, gas pressure in the cylinder, and temperature. All of the data may be important in reconstructing the chain of events resulting in the fatality.

It is vitally important to record this information as soon as possible. Due to limited storage space, some dive computers overwrite the data they have been recording after a period of time. All computers rely on batteries, which may be exhausted and die, resulting in data loss.

When computers are found, the information displayed on the screen should be immediately recorded. The computer should be brought to the surface as soon as possible, even if the body is left for the time being so that it may be sketched or photographed more completely. The computer should be examined by someone competent to do so, and the data downloaded as soon as possible. If low battery indication is noted, it may be desirable to change the battery to preserve evidential data.

The dive buddy's dive computer should also be retained and examined. In at least one case conflicting evidence between the victim's dive computer, the buddy's dive computer, and the statement of the dive buddy were used to place culpability for the incident at least partially on the dive buddy.

Buoyancy compensating device (BCD): Rarely is there a failure of this device that directly affects the outcome of a fatal dive. All too often, however, its absence is a causative factor.

The BCD should be inflated from the diver's air cylinder (or in some cases its own independent cylinder) and be capable of at least lifting at any depth the weight of the lead and other equipment worn by the diver. Small air leaks that may be noticed are usually unimportant, since the main problem experienced with BCDs is the failure of the diver to utilize this important device. On occasion, a victim may be recovered and it may be noted that the BCD was never connected properly or at all; or in the case of borrowed or rented equipment, he may not possess the low-pressure air line or appropriate connector to properly use this piece of equipment. This is, of course, diver error, not equipment malfunction. In short, the BCD should be (a) easy to inflate (usable) and (b) comfortable to wear.

Weight belt and integrated weight systems: It is indeed unfortunate that the misuse of such an inexpensive piece of equipment has directly contributed to so many diving fatalities. Simply stated, divers and diving equipment manufacturers and instruction agencies go to great lengths to provide and explain the use of the quick-release weight system. Despite this, it is almost never used.

When equipment is worn improperly, it cannot function as designed. The body of a scuba diving fatality victim was recovered with the webbing of his weight belt tucked securely into the slot of the lead weights. The chrome quick-release mechanism was forced closed, and the belt could not have been released quickly in the event of an emergency. The red color noticeable in the water is from blood loss due to anthropophagy. As in most diving accidents, this must be classified as diver error, not equipment malfunction. The public safety dive team must observe, record, and reconstruct the accident. Amidst the horror of such accidents there must always be good observation and objective reasoning present in the investigating team.

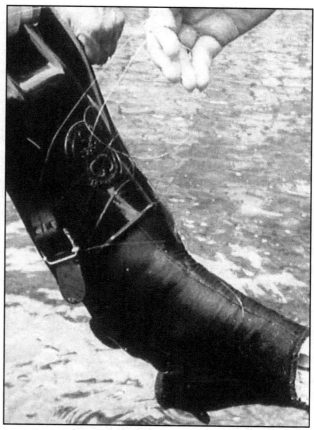

Although not the primary cause of this scuba divng fatality, evidence that the victim had encountered monofilament fishing line gave rise to the suspicion that entanglement may have been one of the many factors leading to ultimate panic. This fatality occurred during the victim's first night dive. He possessed only limited diving experience. The body was located at a depth of slightly more than 100 feet.

There seems to be a trend, particularly among novice divers, to over-weight themselves during most dives while compensating with a large BCD. They frequently add trim weight to the BCD back pockets or on a tank to the point that it adversely affects an ascent should one be needed. Most instruction agencies teach that divers should float comfortably at the surface with no assistance from their BCD. However, due to poor skill levels, most divers tend to add extra weights to their weight system or add excessive weights to the trim system on a BCD in an effort to make the initial descent through the first few feet (~1-2 m) easier. The over-weighted diver, without the judicious use of his BCD or the survival instinct to ditch his weight belt, is at risk. During a panic or near-panic scenario, the inflation of a BCD may be a less-than-instinctive behavior; however,

the ditching of the weight belt is a skill that is an often talked about yet seldom practiced routine. In the case of integrated weight systems on BCDs, the replacement of weight pockets can be an obstacle to practice. If it takes excessive time or energy to replace the pockets on an integrated weight system, many divers will simply not take the time to practice the lifesaving skill. Weight belts and weight pockets cost money, and to ditch a piece of expensive equipment is all too often a rarely considered action.

On occasion, divers will go to great lengths to prevent the loss of their weight belt. Securing the buckle with elastic or tucking the trailing edge under the belt or threading it through a lead weight are only two actions that have been recorded following the recovery of a victim of a diving fatality.

Again, what has been referred to as equipment failure is merely diver error, which is only one step in a series of events leading up to the fatality.

Fins: Rarely is the victim of a diving fatality recovered with a fin missing; however, this has occurred and is worthy of mention. When a fin is missing from the body, the impaired swimming ability is almost assuredly one of the factors that precipitated or led to the fatality. A search for the missing fin is important in order to ascertain whether its loss was due to an improperly installed fin strap or breakage due to deterioration of the rubber. Although these must all be considered as diver error rather than equipment malfunction, at least it will assist in piecing together the chain of events that led up to the accident.

In accidents where the victim was a student whose dressing and equipping was supervised, or where the equipment was rented, the condition of the equipment may be very important in assessing culpability.

Computers: Dive computers provide the investigator with excellent information as to the diver's recent activities while underwater. By having qualified technicians download or otherwise retrieve previously logged information, the investigator can gain important information about the incident. By comparing witness or diving buddy statements to the facts from their dive computers, we can assess if inconsistencies are present in the statements and act accordingly with the investigation.

Other equipment: One of the few pieces of equipment that all too often go unused and are missing from many sport divers' rigs are the submersible pressure gauge (SPG) and the alternate air source (safe second, octopus, etc.). Any fatality investigation that reveals that either the deceased or the survivor failed to be equipped with these two very important pieces of equipment should be carefully reconstructed to determine if either of these pieces of equipment would have changed the outcome of the dive. Divers are often criticized for not wearing (or wearing improperly) depth gauges, watches, knives, decompression meters, back-up knives, compasses, etc., but rarely are these pieces of equipment responsible for a fatality.

Although all gauges and accessories are important and have their places, their mere absence does not necessarily constitute a major safety violation. As can be easily understood, a drowning in 30 feet (9 m) of water cannot be blamed on the lack of a watch or depth gauge; a heart attack on the surface is not the result of a failure to wear a decompression meter, and diving beyond one's physical and mental limitations may place a diver in a fatal scenario from which even the sharpest of dive knives will not extract him.

CYLINDER MARKINGS
Cylinder markings should also be noted during the investigation. These markings should include serial number, cylinder size (in cubic feet or liters), last hydrostatic test date and visual examination date, composition (aluminum, steel, or composite), and any special exemption marking on the tank. Take note of any nitrox markings or stickers. Cylinder valves should be closed (marking the initial position and noting the number of turns it takes to close them) and secured to preserve any gas remaining in the cylinder.

NITROX AND OTHER GASES
Nitrox is a special blend of breathing gas with a higher content of oxygen and a lower content of nitrogen. Its use allows divers to extend their bottom time without incurring mandatory staged decompression obligations, since there is less nitrogen when compared to air. The trade-off is that oxygen partial pressure (PO2) is much higher than with air diving to equivalent depths. As PO2 increases, so does the risk of oxygen toxicity. Prolonged exposure to elevated PO2 may result in adverse effects to the central nervous system with the potential to result in seizures.

Trimix is a combination of helium, oxygen, and nitrogen. It is used in deep diving (generally greater than 130 ft/40 m). Helium is introduced to the breathing mixture to reduce the narcotic effect of nitrogen, since helium does not cause the narcosis. Incorrect use of trimix may result in oxygen toxicity or hypoxia.

Argon is occasionally used to inflate drysuits. Because of gas characteristics, it is believed by many divers to provide superior insulating capability than air. It is not used for respiration. However, if accidentally introduced into a breathing gas cylinder, the effects could be catastrophic.

As investigators we should be checking the gas composition and quality in the cylinder of the deceased diver as well as others in the dive party. SPGs should be checked and dive time also noted for the investigation. Investigators should check for the exact oxygen/nitrogen, oxygen/nitrogen/helium and other trace gases with proper gas-analysis equipment.

REBREATHERS: CLOSED-CIRCUIT SYSTEM

Rebreathers are becoming more popular in the recreational dive community. They are no longer just for military applications. Special considerations need to be applied to these systems. While they all employ the same basic concept of removing carbon dioxide (CO_2) from the breathing loop, they have different sensors that operate in different capacities. It is important to understand how each unit is designed to function. Public safety dive teams should seek assistance from experts in this type of equipment to help with their investigative efforts if this type of equipment is encountered in the course of their duties.

4. THE AUTOPSY

Unfortunately, this portion of the scuba fatality is partially or wholly out of the hands of the investigative dive team. At best, the investigator can only hope for a forensic pathologist who is well-educated in the autopsy of diving-accident victims and is willing to follow established procedures. Perhaps one of the most valuable activities the professional investigative dive team can involve itself with prior to the diving fatality is establishing a good relationship with the local pathologist. When this relationship is well established, territorial boundaries fade, and suggestions from both sides are welcomed.

Rebreathers are becoming increasingly common in recreational diving. Forensic evaluation of fatalities involving this type of equipment is more involved than standard open circuit equipment forensics, and generally requires the assistance of an expert in the rebreather system used by the victim.

The scuba fatality autopsy is a specialized procedure. Although statistics reveal that most commonly the cause of death is "asphyxia due to drowning," air embolism cannot by overlooked, nor should it be. Indeed, every autopsy conducted on a victim of a scuba fatality accident should be a suspected embolism until determined otherwise.

The procedure used to determine "death due to air embolism" is specialized in that great care must be taken in opening the chest cavity so as not to introduce air directly into the coronary arteries. In fact, a fragmented blood line in the coronary arteries is often produced during the early stages of the routine autopsy.

DETERMINING "AIR EMBOLISM" — ARTERIAL GAS EMBOLISM (AGE)

A T-shaped incision is made directly over the heart. This incision can be made so that it may later be extended to effect the normal incision through which most organs may be removed. After the incision is made, the skin and underlying muscles are folded back and the precordial portion of the sternum and ribs are removed. Great care must be taken in the removal of this structure. Careful and complete incisions are necessary, and any pulling to separate these structures should be avoided, since it may result in an aspiration of air into the underlying vessels due to the negative pressure created.

The pericardial sac is then cut anteriorly, and the edges are held back with the use of hemostats. The basin thus formed is filled with water until the heart is submerged. Using a scalpel, the right side of the heart is then incised and the scalpel twisted a few times to ensure that the wound is opened. If air is present in the right side of the heart, bubbles will be seen to rise in the water. If no bubbles are present, the basin formed by the pericardial sac will fill with blood. In cases of massive embolism, it is possible to witness almost no blood escaping.

The collecting and subsequent analyzing of these bubbles can be done by inverting a water-filled graduated glass cylinder with a volume of at least 300 cc into the basin then sealing it off with a rubber stopper after as much gas as possible has been collected. This last step, while desirable, is not necessarily as important as the demonstration of air within the heart itself, since the gas will nearly always analyze as nitrogen with varying amounts of oxygen. The gas remaining in the circulatory system after most air embolism or even decompression-sickness-related deaths will often possess less oxygen than the atmosphere, merely because most of the oxygen is absorbed by the surrounding tissues. The presence of the gas, carefully demonstrated, is more important in the diagnosis than its makeup.

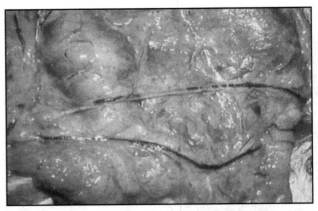

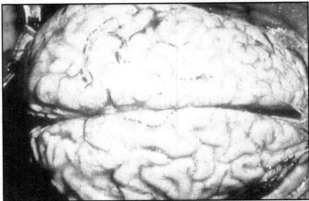

Arterial gas embolism (AGE) is often demonstrated during the autopsy in what is commonly called the boxcar effect. Like the cars of a train, the blood and gas contained in major arteries (and veins) often appears intermittently. Aggressive cardiopulmonary resuscitation attempts during the rescue phase may evenly distribute these bubbles throughout the circulatory system — venous as well as arterial. Since putrefaction also produces gases within the circulatory system, this boxcar effect can often be demonstrated on cadavers that had never been exposed to a hyperbaric environment. As with any pathological finding, determination as to the cause of death is usually a culmination of many observations. In this case, death was found to be an air embolism. This finding was corroborated with other factors, including a panic ascent, loss of consciousness within two minutes after surfacing, and the victim's last words: "I've done it… I'm dead."

DETERMINING PNEUMOTHORAX

Pneumothorax occurs during the "burst lung syndrome" and may be present or absent even though an air embolism can be demonstrated. Pneumothorax occurs when sufficient damage is done to one or both lungs to allow air to escape directly into the chest cavity between the lungs and the chest wall. Pure pneumothorax is rare because of the massive damage done during the lung overexpansion. More commonly a hemopneumothorax is present. A hemopneumothorax is a condition in which both blood (hemo) and air (pneumo) are found in the chest cavity outside the lungs.

The demonstration of a hemopneumothorax is rather simple and should be carried out prior to removal of the breastplate. To test for the presence of a hemopneumothorax, the side of the skin from the chest (after the initial T incision) is reflected and cut to form a side pocket or pouch. This pouch is then filled with water, and, using a scalpel, an incision is made through an intercostal space (between the ribs) into the costodiaphragmatic sinus (the space above the diaphragm and exterior to the lungs). Several of these incisions should be made, with the scalpel twisted each time to ensure that the wound thus formed is clearly opened. Escape of a gas from these wounds clearly demonstrates a pneumothorax. The escape of a blood froth from the incision demonstrates a hemopneumothorax.

As in the procedure for determining air embolism, an inverted graduated cylinder may be used to collect and measure the volume of this gas, but this procedure is considered by some to be superfluous.

THE USE OF X-RAYS, MRIs, AND CT SCANS:

The use of an X-ray prior to the autopsy has been very successful in revealing the presence of a pneumothorax in victims of diving fatalities. Its use, however, in demonstrating embolism is questionable at best due to the small volume of air remaining in the heart or circulatory system. When an X-ray diagnosis is to be attempted, the victim's body should be X-rayed at different angles to the chest cavity. CT scans and other technology have been used recently to determine causes of death, and these technologies should not be overlooked when applicable.

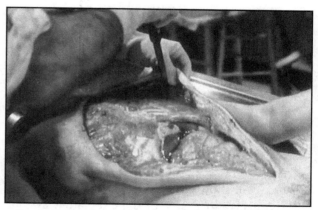

Determination of pneumothorax (air in the chest cavity) is often made by forming a side pocket or pouch; thus formed, it is filled with water and a knife is inserted through the intercostal space (between the ribs). Once the knife has been inserted into the costodiaphragmatic sinus and twisted, bubbles may be seen to exit the incision.

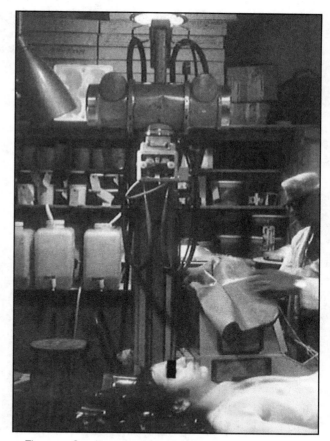

The use of radiography (X-rays) is common in the determination of cause of death. An X-ray may reveal fluid-filled lungs (drowning) and large air spaces (pneumothorax). Many pathologists routinely X-ray the chest prior to incision.

ARTERIAL VS. VENOUS GAS EMBOLI

The question of determining whether death is due to air embolism or decompression sickness becomes almost academic in the presence of lung damage due to overexpansion. There are authorities, however, who have stated that an autopsy of a victim who has suffered death due to air embolism will display gas emboli on the arterial side of the circulatory system, while death due to decompression sickness will show gas bubble formation chiefly on the venous (return) side of the circulatory system. Clearly this is not so, and the reason is simple: In accidents involving death due to decompression sickness, the onset of death is rarely sudden, and the bubbles formed from nitrogen coming out of solution in the tissues and traveling through the circulatory system have had ample time to establish themselves equally on both the venous and arterial sides of the circulatory system.

In cases involving a sudden collapse and subsequent death due to air embolism, it may be true that the bubbles are primarily noted on the arterial side of the circulatory system, but it has been documented that effective application of cardiopulmonary resuscitation (CPR) is capable of circulating these bubbles evenly throughout the brain and other sites in the body.

Conclusion

Since it is not uncommon for diving accidents to be precipitated by health problems such as heart attack, emphysema, etc., a complete autopsy is in order. The medical/forensic autopsy is an important step in any fatality investigation, but it should be understood by the professional investigative dive team that it is only one of many steps. Findings made during any step or part of the investigation are strongest when they are corroborated by independent findings that may come to light during a different phase of the investigation.

The autopsy should also include other routine procedures, including the appropriate investigation into the presence of alcohol or other drugs.

5. FORMULATING A CONCLUSION — A REASON FOR THE FATALITY

To the lay person, scuba fatalities present a contradiction in logic. There is often no specific overriding reason behind the accident; indeed, it seems as if it were a combination of events and errors so small that if considered individually, they would hardly be worthy of mention. To effectively investigate a diving accident, this truth must be understood. Rarely does a fatality result from one safety violation, one piece of diving equipment poorly worn or maintained, or even one lapse in good, effective, and thorough training.

The following common details are often observed when recovering the body of a victim of a scuba fatality. When these facts are considered individually, without understanding just what had happened to lead up to them, they present the lay person with an unsolvable puzzle. It is the responsibility of the trained and experienced investigator to unravel this mystery and in doing so make sense out of the accident so that coroner's and other courts will not come to faulty decisions. Making sense out of an otherwise meaningless accident will also help the next of kin to accept what has happened, without being haunted by a lifetime of unanswered questions.

Common Observations — Open-Water Scuba Fatalities

1. Upon recovery of the victim, it is noted that there is sufficient air remaining in the cylinder for a safe and leisurely ascent. In most cases, tank pressure exceeds 700 psi (48 bar).

2. At depth (i.e., the depth where the victim was located), there is very little air in the buoyancy compensating device.

3. The weight belt is still worn, and no obvious attempt has been made to ditch the weights.

4. The mechanical inspection of the regulator fails to reveal any major problems. While it may have been operating at less than manufacturer's optimum specifications, it is still within "safe" limits.

5. Air analysis of the contents of the scuba cylinder is within normal "safe" limits. The water content (dew point, relative/absolute humidity, etc.) may be high, but breathing moist air would hardly seem to be a contributing factor.

6. Equipment worn by the victim was worn correctly and in fair to good overall condition. The equipment may have not been the best available, but if critiqued objectively and fairly, it was certainly sufficient.

7. Diving conditions were less than excellent but seemingly acceptable. Often waves, current, reduced visibility or depths in the vicinity of 100 feet (30 m) are encountered, but when considered carefully, the conditions would not normally be referred to as unsafe.

8. Buddy contact during the dive was (reportedly) good, with the survivor usually commenting that "we only lost contact for a few seconds/minutes before I noticed him missing."

9. No major safety violations were cited that could be directly attributed to the cause of the accident. Although often minor safety violations are documented, these are common occurrences that are usually resolved easily during most dives.

10. The victim was found with his regulator mouthpiece lying beside him.

EXPLAINING THE OPEN-WATER SCUBA FATALITY

Since most of the previously listed conditions will present themselves either as isolated observations or together in the majority of scuba fatalities, an attempt will be made to explain them point by point.

Sufficient air remaining in the cylinder: Most diving accidents are precipitated by a panic episode. Since panic is usually recognized as a "sudden and unreasoning fear," the usual reaction to panic is an instinctive one. Instincts invariably override consciously learned responses. When confronted with danger or panic, rapid breathing is an instinctive response. To facilitate this response, the mouth is usually opened wide. This is contrary to survival when using a normal regulator/mouthpiece, where in order to ensure an uninterrupted air supply, the individual must bite down on the mouthpiece and keep the lips tightly closed to forbid the entry of water into the mouth. Compounding this problem is the fact that the nose, which is a secondary air passage, cannot be used because of the face mask.

Certainly, most regulators presently in use will supply a sufficient flow of air, but the physical action that must be maintained to use them is contrary to what instinct would drive the panic-stricken diver to do.

Insufficient air in the BCD: In most cases, this observation is very important. Unless the diver was involved in an activity where it was desirable to remain stationary on the bottom, the judicious use of a BCD helps to inhibit the stirring up of bottom silt (hence reducing visibility) and allows for a more relaxed dive. Certainly there may be little or no need for air to be injected into a BCD during a very shallow dive, but in general a diver who fails to utilize his BCD is placing himself in a position of hard work and reduced visibility. This is not a desirable option — especially for the novice diver.

Weight system still worn: While this is not meant as a criticism of current teaching standards, it is unfortunate that the practice of self-rescue (weight system ditching) is always taught but rarely practiced. Many studies have shown that the ditching of a weight system, even while wearing a drysuit, does not result in an uncontrollable ascent. Certainly, an ascent rate of 30 feet per minute (the standard rate recommended to avoid decompression sickness) may be exceeded, but decompression sickness is usually easier to treat than drowning. Research is needed to reveal reasons why divers found on the bottom are still wearing their weight.

Over time, advancements have been made to the BCD with integrated weight systems that allow the diver a more comfortable fit. However, divers often are not afforded enough time practicing ditching procedures because the weight pouches are expensive. In any case, it is important to be familiar with these systems and note which system was being used by the diver. Even if the practice of weight ditching were taught and practiced, most divers are hesitant to throw off a piece of costly diving equipment. Bearing this in mind, even the most rigorous training in weight ditching may not result in more divers ditching their weights to avoid drowning. In addition to this, it is worthy of mention that in a panic situation, the overriding instinct would be to swim for the surface, not shed equipment, the latter being a conscious and deliberate action.

Mechanical inspection of regulator — satisfactory: Most diving fatalities result in a series of small human errors. Major equipment malfunction is rarely cited as the only, or even the chief, cause of scuba fatalities.

Air analysis of cylinder—normal: In general, the sport diving community has done an excellent job of policing itself. Air from most compressors either meets or exceeds the industry standard. Rarely is carbon monoxide contamination present, and oxygen depletion from internal cylinder corrosion has been documented only in the rarest of cases.

Equipment worn correctly and in good condition: Most sport divers are both proud and safety-conscious. When errors are made, they are usually minor. While in many diving communities the wearing of a diver's knife on the outside of the leg is considered a major violation, it has rarely resulted in a documented fatality. Indeed, the fact that the diving community has become so demanding on such minor details reveals the fact that in general sport diving is one of the most safety-conscious sports in existence. In cases where the victim is recovered and equipment is worn incorrectly or even missing, an objective evaluation should be made to determine whether that was one of the causes of the accident. Certainly, a victim who was found not wearing a knife could not be blamed for violating a safety dictate if entanglement was not a factor in the accident.

Diving conditions acceptable but less than good: Quite often the overall conditions of the environment play a significant role in the sport diving fatality. Waves may arise that were not present during the beginning of the dive, which make the swim to shore arduous. An unexpected

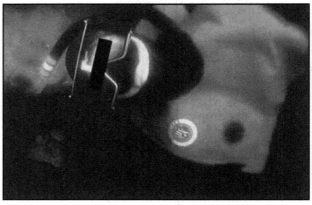

It is common to find the victim of a scuba fatality accident without his mouthpiece in place. Not only do all muscles typically relax after death, but the agonal-gasp phase usually involves the victim's mouth being open widely and then relaxing as consciousness is lost.

current may be encountered, especially in ocean diving. The visibility of a clear lake may be reduced to near-zero by an overactive, heavily weighted diver. Similarly, an over-weighted diver or a diver who does not utilize his BCD judiciously may find himself in deeper water than originally planned. The increasing depth as he descends a steep incline may itself precipitate a panic response.

When diving conditions are considered in the formulating of a reason to explain the fatality, the whole picture must be considered. Chiefly, two questions must be answered: (a) Did the conditions change unexpectedly during the dive, or (b) was the victim familiar with the diving conditions present during the accident?

Buddy contact was momentarily lost: The comment that "they were there when I last looked" indicates that buddy contact was not a constant condition of the dive. To be effective, buddy contact between two divers must be a continuous, ongoing condition of the dive. Buddy contact that is checked every few minutes is, at best, poor. Unfortunately, this interrupted and noncontinuous method of buddy diving is more common than many divers would like to admit. It is difficult to criticize the survivor of poor diving practices in cases such as this because his position is one of a lost friend. This less-than-optimum buddy system is an error of omission that is common among most sport divers. In short, human error is probably more to blame than negligence.

No major safety violations documented: In the vast majority of diving fatalities, there are no major safety violations reported as a result of the investigation. Instead, when carefully investigated, many diving accidents are found to be a result of small errors and incidents that when combined placed the victim in a position he was unable to escape from. To formulate such a conclusion requires a slow, meticulous, and objective investigation of all aspects of the fatal dive.

Victim found with mouthpiece removed: As has been already discussed, basic instincts force the mouth open to facilitate easier breathing when panic has been established. Even in cases where this has not occurred, drowning is usually followed by primary muscle flaccidity. This relaxing of all muscles allows the regulator or mouthpiece to easily drop from the mouth of the victim.

THE ACCIDENT SCENARIO — CONCLUSION

Perhaps the lay person will never understand why in most open-water scuba sport diving fatalities the following are true.

(a) The victim is found on the bottom.

(b) There is little or no air in the BCD.

(c) The diver's weight belt is still on, even though it was equipped with a quick-release buckle.

(d) All equipment passed a mechanical inspection.

(e) The contents of the cylinder analyzed as normal atmospheric air.

(f) All equipment worn by the victim was worn properly and in good condition.

(g) Diving conditions (the environment) were, at the very worst, acceptable.

(h) Buddy contact was maintained somewhat throughout the dive.

(i) There were no major safety violations that could be directly linked to the fatality.

It cannot be overemphasized that most sport scuba fatalities are the result of several events. Each event,

when encountered on an individual basis, could easily be overcome, but together they presented a scenario that put the diver in a position from which he could not escape. Panic followed, and ultimately death was the result.

SCUBA FATALITIES — ICE AND PENETRATION DIVING

Each year more and more sport divers are becoming involved in diving beneath an ice canopy. Under-the-ice diving is a specialized activity that is inherently safe but carries with it its own specific set of guidelines and safety rules that must be followed. Similarly, penetration diving, whether it be into wrecks or caves, carries with it its own set of hazards that must be taken into account. Only proper professional training and safe diving procedures will adequately protect the sport diver from the hazards involved in these specialized activities. Accidents most commonly occur when the diver has attempted one of these highly specialized activities without receiving appropriate training.

When a victim of an ice dive or penetration dive is recovered, in most cases he will be found with a depleted air supply and (especially in the case of an ice-diving fatality) will likely have ditched his weight belt.

Since many of these fatalities involve divers who have not been formally trained in this activity, they are generally quite easy to explain, provided the professional investigative dive team has had the necessary training to understand what should have been done.

Very few public safety dive teams have the training, specialized equipment, and experience necessary to perform investigations or body recoveries from submerged cave systems and other overhead environments. Merely being part of a public safety dive team, even a very experienced one, does not prepare you to work in these environments. There have been multiple incidents where public safety dive teams have attempted to perform such work, with the result that the entire team perished trying to perform their duty. In one memorable incident, the back-up team also died trying to recover the bodies of the first team they sent in.

Even if the overhead environment is a virtual one (for example, the ceiling imposed on divers by incurring a decompression obligation), public safety dive teams should NOT attempt the dive if they do not have specific training for that environment. So if your team is faced with such a challenge, what should you do?

The International Underwater Cave Rescue and Recovery (IUCRR) team was formed for this purpose. The original team was formed in 1982 in Florida, where there are hundreds of springs and underwater cave systems. There were so many diving fatalities in these systems for which local public safety dive teams were unprepared to enter that Henry Nicholson, then with the sheriff's department in Jacksonville, Duval County, Florida, formed a team comprised of volunteers from throughout the state that had the specialized knowledge to dive these locations safely. He developed a training program to further prepare them in the basics of underwater investigations and recoveries.

Since 1999 the IUCRR has grown to an international all-volunteer, not-for-profit public service and educational organization registered in the state of Florida. Their mission is to support all public safety agencies and work within their incident command system in the rescue and/or recovery of victims in an underwater-overhead environment (environments partially or fully underwater with an overhead obstruction such as caves, caverns, mine shafts, etc., and includes virtual overhead environments as well). They work under the direction of the local law enforcement group that seeks their assistance and provides these services at no charge. Since the IUCRR is a virtual organization, there is no office address or phone number. However, if you need to contact them to assist with a recovery, you can use the following methods: send an email to iucrr-info@cavediver.com or call the current director (as of the time of this writing) at +1-863-686-8285.

ON THE SCENE: SCUBA FATALITY

"Reasons for Scuba Fatality Unclear": The newspaper headlines and subsequent articles could not easily place blame or give a simple reason for a sport diving accident, so it was labeled a mystery.

The Scenario

Two newly certified sport divers had been diving in a nearby lake. This was their (approximately) 20th dive and their second descent in this location. The day of the dive saw ideal weather conditions, although the wind was causing some wave action to begin during their swim out from shore.

At a distance of approximately 100 feet (30 m) from shore, both divers descended into the clear water and settled to the bottom at a depth of 30 feet (9 m). The visibility was in excess of 20 feet (6 m), and after establishing buddy contact, they set about exploring the bottom. Approximately 50 feet (15 m) from their location was the edge of a sheer drop-off. Their intention was to explore the caves they had heard about but not to penetrate more than their own body length.

Once the pair of divers had reached the edge of this underwater cliff, they were able to look over the edge and see the mouth of the first cave. Together they swam over the edge of the drop-off. The survivor related in his statement to the police that he glided down to the cave entrance with his buddy following. Once at the entrance, he turned on his diving light, checked his tank pressure (he stated that his gauge read 1800 psi/122 bar), then entered the cave "up to my knees."

The large number of catfish residing within the first few feet (first meter) of the entrance caught his attention, and after only a minute or two, he turned to his buddy to signal that it was his turn. His diving partner had disappeared. Thinking that his buddy was exploring another catfish cave, he went looking for his partner. The survivor went on to explain that there was a current that tried to "drag me down into deeper water," but he managed to make it back up the cliff in a hand-over-hand fashion.

The local public safety dive team located the missing diver the following day directly below the catfish caves in 105 feet (32 m) of water. His equipment was in place and worn correctly, with one exception. His face mask was partially dislodged. His tank pressure read 1400 psi (95 bar).

What Happened?

As can be seen, this scenario follows a pattern. The swim out was tiring, and once having reached a reasonable distance from shore, the pair descended quickly to the bottom, breathing from their scuba tanks. Neither diver added any air into their BCDs; negatively buoyant, they began their further descent to the mouth of the first cave — a depth of 50 feet (15 m). As the first diver entered the cave, his partner (probably) followed close behind.

What happened next is subject to conjecture, but it is entirely possible that in his attempt to close the distance between himself and the wall (or his buddy), the victim's face was kicked and his mask dislodged. Already negatively buoyant and with a flooded mask, the victim's descent down the vertical rock wall was swift. In the survivor's few minutes of distraction while he was observing the interior of the cave, his diving partner descended uncontrollably, gulping water in through his nose, negatively buoyant, and drowned. His last efforts were likely an unsuccessful attempt at replacing his mask, grasping for the safety of the rock wall or attempting to swim to the surface, heavily weighted, having difficulty breathing and partially blind.

Conclusion

It would be easy to fault the survivor for his loss of buddy contact, the deceased for failing to utilize his BCD, or both for a combination of small errors.

The objective investigator, however, should merely collect facts and report his findings in a clear and objective manner. Opinions may be given during subsequent judicial hearings or at the end of the actual report, but in coming to any conclusions this should be remembered: There is a great difference between explaining an accident and finding fault.

The investigator is responsible for the explanation only. In all scuba fatality investigations, it must be remembered that although errors may have been made, there always exists the ultimate explanation that it was indeed an accident.

6. THE SCUBA FATALITY INVESTIGATION REPORT

The culmination of any investigation is a clear, concise, objective, court-ready report.

The following is provided as a guideline for the investigator. No single guideline can serve as a fill-in-the-blanks form for all investigations, and that is not the intention of this suggested guideline. The sole purpose of the guideline provided here is to provide the investigative dive team with a place to start and a route to follow.

This guide should be used in concert with the Body Recovery Report and Supplemental Underwater Recovery Report — Postmortem Observations.

SCUBA FATALITY INVESTIGATION REPORT

Case Heading: _____

_____

Dive Team Members: _____

_____

_____

_____

_____

Report Submitted By: _____

In Charge Of Investigation _____

Name _____

Address _____

_____

_____

Phone Number _____

RÉSUMÉ/SUMMARY

Enter a brief account of the events leading up to the diving accident. Do not repeat details that will appear on following pages.

_____

_____

_____

_____

_____

_____

_____

_____

_____

_____

_____

_____

_____

_____

_____

_____

_____

_____

_____

_____

_____

ACCIDENT DETAILS

Location _____

Diving Activities _____

_____

Depth _____

Visibility _____

Current _____

Weather _____

_____

Dive: ❑ Boat ❑ shore

❑ Decompression or ❑ No decompression

Bottom conditions _____

_____

Accident occurred during the following (check)

    ❑ ascent   ❑ descent   ❑ on surface   ❑ at depth

    ❑ on bottom   ❑ beginning   ❑ middle   ❑ end (of dive)

Misc. Comments _____

_____

_____

_____

_____

_____

_____

_____

_____

CASUALTY REPORT

Casualty # _____

Name _____ Age _____

Address _____

_____

Diving experience _____

_____

Recent medical _____

_____

Health _____

DIVING EQUIPMENT

Snorkel: ☐ yes ☐ no

☐ Integrated or ☐ Conventional weight belt _____ pounds/kilograms

Flotation device (Type and amount of air in it when found):_____

_____

Regulator (type) _____

Apparent condition _____

Inspection results _____

_____

_____

Octopus (alternate air source) _____

_____

_____

DIVING EQUIPMENT
Continued

Submersible pressure gauge _____

Tank Valve: ❑ "J"   ❑ "K"   ❑ "Y"   ❑ "H"   ❑ Manifold

Inspection results_____

_____

Tank (diving cylinders) check one: ❑ Steel   ❑ Aluminum   ❑   Composite

Size _____ cu ft/liters   Air pressure when found_____ psi / bar

Manufacturer _____ Serial number _____

Last hydro examination date ___/_____   Last visual examination date   / _____

Nitrox or other special gas label present   ❑ Yes   ❑ No

If yes, describe: _____

_____

_____

Watch: ❑ Yes ❑ No   Depth gauge: ❑ Yes ❑ No

Knife: ❑ Yes ❑ No   Worn properly: ❑ Yes ❑ No

Location on victim:

Additional cutting devices (i.e., EMT shears, wire cutters, line cutters): ❑ Yes   ❑ No

Location on victim _____

Autopsy results _____

_____

Pathologist (hospital name) _____

_____

Comments _____

_____

GAS ANALYSIS

Dive computer recovered: ❑   Yes   ❑   No

Type _____ Where worn _____

Compare date/time settings to standard reference timepiece. _____

_____

Record any information is/was displaying _____

_____

_____

Note computer set-up options (nitrox mix, fresh/salt water, OC/CCR, etc.) _____

_____

Observations _____

_____

Lab sent to _____

Download dive profile and other data. Attach results to this report.

SURVIVOR REPORT

Survivor # _____

Name _____ Age _____

Address _____

_____

_____

Phone number _____

Diving experience _____

_____

Apparent health _____

_____

Was survivor deceased's buddy?	❑ Yes	❑ No
Did survivor recover body?	❑ Yes	❑ No
Artificial resuscitation/ cardiopulmonary resuscitation attempted?	❑ Yes	❑ No
Victim given oxygen?	❑ Yes	❑ No

EQUIPMENT
Regulator (type)_____

Octopus (alternate air source) _____

Submersible pressure gauge _____

Tank Valve: ❑ "J" ❑ "K" ❑ "Y" ❑ "H"   ❑ Manifold

Tank (diving cylinders) check one: ❑ Steel   ❑ Aluminum   ❑ Composite

Size _____ Cu ft/liters   Air pressure when found_____ psi / bar

Gas analysis:_____

Flotation device: ❑ Yes ❑ No   Type _____

SURVIVOR REPORT
Continued

Prior alcohol/drug use

_____

_____

_____

_____

ADDITIONAL REMARKS

To be recorded by dive team member — including your opinion as to the major contributing reason for this fatality.

_____

_____

_____

_____

_____

_____

_____

_____

_____

_____

_____

_____

_____

_____

_____

_____

INQUEST

Inquest held   ☐ Yes   ☐ No

Coroner _____

Location _____

_____

_____

Coroner's verdict _____

_____

_____

_____

_____

_____

_____

Jury recommendations _____

_____

_____

_____

_____

_____

_____

_____

_____

_____

_____

NOTES

FIREARMS RECOVERY

8

It is indeed unfortunate that the proper procedure for firearms recovery and, in particular, preservation is often learned too late. In all too many cases public safety dive teams have set out to find a gun that was used in a serious crime, and through days of incredibly hard work they succeed in retrieving a firearm that, upon presentation to the forensic laboratory, is virtually useless as evidence.

Once a mistake of this magnitude has occurred, the dive team has not only failed to act as professionals but also has lost credibility with their sponsoring agency. Credibility, once lost, is all too often never regained.

The reason for adequate education in firearms recovery and preservation is then twofold. First, firearms are tools that are often used in the most serious of crimes — homicide. The firearm, therefore, in most cases will have value as evidence. Second, once the gun is submerged in water, it begins to lose its value as evidence. How quickly it is recovered and how carefully it is treated will determine, ultimately, its value as evidence in any court of law or criminal investigation. Of course, the first step is to locate the gun that has been thrown into the water. Unless the weapon is first located, any further consideration is academic.

THE FIRST CONSIDERATION — BUOYANCY

As simple and obvious as it may sound, one of the first facts to ascertain prior to beginning a search for a gun that has been thrown into the water is to determine whether or not that specific gun will float or sink. Admittedly, most firearms sink readily, but there are many small-bore rifles possessing a large wooden stock that will float when placed in water. Most large-caliber rifles will sink when placed in water, and nearly all handguns will sink quickly. The future may hold some surprises for the investigative dive team, however, with the introduction

of more wood and plastics into the manufacture of handguns.

There is only one technique to positively determine whether or not a specific weapon will float or sink. An identical firearm must be placed in the water and the results observed. Obviously, many types of firearms are known to be dense enough to sink, but when any doubt exists, actual immersion of an identical weapon may be necessary prior to commencing the search. In many instances, when a gun is thrown into a swiftly moving river, the buoyancy of the firearm is critical. Even when it is known that a specific gun will sink, it may possess enough buoyancy to be easily moved along with the river current.

When testing the buoyancy of the missing weapon with an identical weapon, it will be necessary to secure its retrieval with a thin, strong line. The line should be strong enough to avoid loss of the second weapon but thin enough to not significantly increase the water drag on the gun. Safety precautions must be taken to ensure that the dive team will only be searching for one gun

The question remains: Where will it be possible to obtain an identical gun for trial purposes? Certainly, retail stores will not lend a weapon for such a purpose, and few gun collectors/owners would allow their firearms to be used in such a manner, but many police departments have incredibly large collections of seized firearms — many awaiting auction or destruction. Police department "exhibit" or "property" departments can furnish such a weapon if the request is made properly through the right channels.

When it is determined that the gun being searched for will remain stationary on the bottom of a river bed, the only other question to answer is: "Will it be covered up?" This question can be answered by simply placing small lengths of pipe at various locations on the river bed. After an appropriate

While the buoyancy of any gun thrown into the water should always be considered, most firearms will sink quickly. This .12-gauge shotgun that was used in an assault luckily came to rest partially on a submerged log. Otherwise, the bottom silt of the lake would have easily covered the shotgun, and the use of the water underwater metal detectors would have to be considered.

This sawed-off shotgun that was used in an armed robbery, kidnapping, and extortion was almost totally covered with smooth rocks that were constantly being transported along the bottom of the river. The gun had been previously missed but then located during an intensive subsequent search by the investigative dive team.

length of time, they may be checked to see if river-bottom material has moved and concealed the metal objects. Any current in excess of 3 miles per hour (4.8 kpm) is quite capable of moving baseball-sized rocks, along with sand and gravel. Many guns have gone undetected during the primary search phase of a mission simply because they were concealed by a moving river bottom.

On the Scene:
Divers Uncover Evidence — Barely

Two adult males, after a foiled kidnap attempt, threw their weapons in a shallow, clear river. The weapons, a sawed-off shotgun and a sawed-off .22-caliber rifle, were thrown from a bridge approximately 15 feet (4.6 m) above water level. The water depth was only 3 to 4 feet (90-120 cm) deep, and the bottom of the river could easily be seen from the bridge.

Within a week the public safety dive team was called to search for the weapons. The river, although clear and shallow, was deep enough to require a walking search by suited divers wearing masks and snorkels only. The current, which was approximately 3 miles per hour (4.8 kph), was too fast for a searcher wearing chest-high rubber waders. Because of the possibility of losing their footing, all searchers wore rubber diver's suits and used safety lines, which also served as search lines for pendulum-arc search patterns. The bridge served as the base for the arc pattern.

The river was searched for a full day with negative results. Neither gun was found, and it was believed the current may have carried them downstream. Similar weapons were brought in and tested. They sank quickly and remained stable on the bottom. It could not be understood why the original firearms could not be found under such ideal conditions.

After the first day's negative search, several 1-ounce (28-g) steel weights were placed on the river bottom. Each was marked with a short length of fluorescent plastic surveyor's tape. The first day's search area was clearly defined for the divers who would be continuing the search several days later.

With the arrival of the second dive team, it was noticed immediately that all the markers previously placed in the river were missing. A closer inspection by the dive team revealed that they had not been carried down river as previously expected but were merely covered up; the bottom of the river was moving. The river bottom was covered with smooth round rocks that were 1-5 inches (25-127 mm) in diameter.

Since the first dive team had searched the most likely area (the area where the suspects admitted throwing in the firearms), it was decided to once again search this area carefully — if necessary, repeating the search with metal detectors. The latter was not necessary; within three hours a portion of the shotgun was seen in only 30 inches (76 cm) of water.

Eventually both guns were recovered from the river bottom, photographed and preserved. The diver's knife that appears on the snow beside the shotgun and the .22-caliber rifle is for scale only. It provides perspective as to the size of the guns.

It was almost completely covered with the smooth rocks that lined the river bottom. A further search revealed the .22-caliber rifle, which was only a short distance away. Both firearms had been passed over by the first dive team.

THE FIREARM AS EVIDENCE

While it is easy to understand that a firearm is indeed evidence, what must not be overlooked is that a gun is a composite of parts and pieces, each capable of offering different types of evidence. Because of this fact, each part may have to be handled differently once the gun has been located. To understand this procedure, the firearm will be discussed under the heading of each individual component. Dive teams are encouraged to seek advice from their forensic firearms testing laboratory whenever a search for a gun is being planned, since different testing procedures require different handling procedures.

The Barrel

The barrel of a gun is considered to be the portion of the firearm that the bullet travels through after its explosive discharge from the shell casing. The inside of the barrel is marked with a spiral pattern, known as rifling. (This is the origin of the term rifle; however, rifling is also present in pistols.) These raised and lowered configurations, or "hills" and "valleys," forming the spiral grooving of the inside of the barrel are referred to as lands and grooves. The lands are the raised portion of the spiral, and the grooves are the valleys. The purpose of this spiral is to set the bullet spinning;

this spinning action gives the bullet many of the properties of a gyroscope and greatly increases the accuracy of the shot.

Lands and grooves are found in all modern firearms of virtually all makes and caliber. (Shotguns are an exception to this rule, since they are not normally used to fire a slug. A shotgun is said to be a smoothbore weapon.) The physical dimensions of these spiral configurations will vary from firearm to firearm, manufacturer to manufacturer, and even within various models of similar weapons.

Perhaps the most well-known forensic value of recovering a firearm is in what is termed the rifling marks or rifling characteristics. As a bullet, whether it be bare lead or a metal-jacketed lead slug, travels down the barrel, the spiral lands and grooves on the inside of the barrel score their own characteristic markings onto the slug. These marks, which are left on the slug, are the rifling marks. They are characteristic of only one firearm in the world, the one that fired the slug. Hence, when a slug or bullet fragment is located at the scene of a crime and enough of it remains intact to exhibit sufficient rifling marks, it becomes a valuable piece of evidence. If the suspect weapon is located and test-fired, the test bullet may be microscopically compared to the slug found at the scene of the crime. If, upon comparison, marks on both slugs match, the gun that fired the test slug has now been proven to be the same weapon that fired the slug found at the crime scene.

One fact that is often not understood, however, is that an intact slug does not have to be used for comparison purposes. Depending on the specific number of positive comparison points required for court proof, often a slug fragment or distorted slug found at the scene of a crime may offer enough characteristics for positive comparison. The number of specific comparison points required to positively match two slugs may vary from court jurisdiction to jurisdiction or police department to department, depending on local policy or previous court rulings.

Because the spiral markings inside the barrel of a gun are so important, nothing should ever be placed in the barrel of a firearm that has been recovered.

Any scratches, even microscopic marks, that will be left on any test bullet fired from the suspect gun may render it useless as evidence. All attempts possible should be made to retain the original rifling characteristics of the firearm that has been recovered. Even the temptation to pick up a handgun using a soft wooden pencil inserted down the barrel should be avoided at all costs.

This point deserves repeating: Because the spiral markings inside the barrel of any firearm are so important, nothing should ever be placed in the barrel of a gun that has been recovered. Any scratches, even microscopic marks, may alter the configuration of the rifling marks that will be transferred to any subsequently fired bullet.

The Ejection Mechanism
Many firearms (both rifles and handguns) are equipped to eject the spent cartridge after the round has been fired. The mechanics of the ejection mechanism may vary from firearm to firearm, but all mechanisms are dependent on pressure being exerted on the shell casing to forcibly eject it from the firing chamber. The mechanism of the ejector is not an important factor, but whether the spent cartridge is ejected automatically or manually, considerable force must be exerted on the casing in order to force it from the gun.

When the spent casing is ejected, there are marks left on it by the ejection mechanism. Often these marks are not obvious but will become apparent under microscopic examination. Because of characteristic wear of the ejection mechanism, its manufacture (machine) marks, etc., each ejector mechanism will leave its own fingerprint on the shell casing. This fingerprint that is left on the casing can often be matched to the mechanism that ejected the shell casing. When this is done using comparative microscopic examination, an ejected shell casing found at the scene of the crime can often be matched conclusively to the firearm from which it came.

The mark left by such an ejection mechanism may not be obvious to the casual observer. For this reason, all casings located at the site of a recovery must be handled carefully. In addition to considering the handling of any found shell casings (spent cartridges), the investigator should also be aware of the fact that the spent cartridge is only half of the evidence necessary to effect a physical match. Of course, the other half in this equation is the ejection mechanism of the firearm. For this reason, it is imperative that all firearms be recovered and preserved in total for further examination by a qualified firearms technician.

The Firing Pin
As well as the physical, often microscopic, marks left on a shell casing by the ejection mechanism, a distinctive mark is produced on a shell casing by the firing pin of a gun. This is true whether the firing mechanism is referred to as a center-fire (where the firing pin strikes the casing in the center) or a rim-fire (where the firing pin strikes the rim of the shell casing). Either way, an indentation will be left on the shell casing.

Under microscopic examination, this indentation may be seen to contain specific characteristics or marks that can be physically matched to the firing pin and consequently the gun that fired the bullet.

RECOVERY AND PRESERVATION OF AMMUNITION
Bullets or even bullet fragments possess rifling marks characteristic of the barrel they traveled through. They should be handled very carefully to avoid any further marking or damage of the already-present rifling marks (lead is soft). Ideally, bullets should never come in contact with any hard or abrasive substance. Bullets should be recovered using either bare hands or rubber diving gloves. If the exterior of the rubber diving glove is covered with nylon, extreme care must be taken and the bullets handled very lightly. Once lifted from their resting place, they should be immediately placed in plastic bags or containers while still underwater. Each bullet should be placed in its own container to avoid further damage from friction or collision inside the container. Once removed from the water, each bullet should be carefully and gently rinsed in fresh clean water, allowed to air dry, and then packed in a soft material for transportation. The bullets should never be dried by means of a cloth or towel.

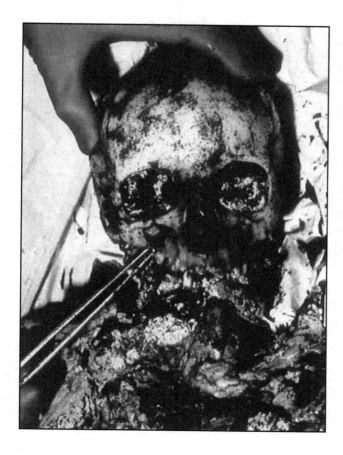

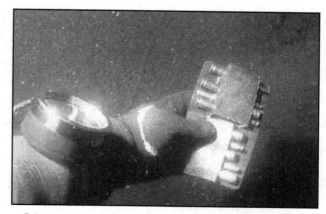

Even ammunition should be recovered carefully. This ammunition that was thrown into the water by the suspect came from the same box as the ammunition used in a homicide.

The slug being removed from the skull of this homicide victim was from the ammunition box photographed above. The ammunition and the body were recovered miles apart. Note the pathologist is using metal tweezers to remove the slug. This is poor practice, since any marks inflicted on the slug may interfere with ballistics comparison. Investigators should be present during autopsies to receive exhibits and to assist the pathologist.

Shell casings may possess three types of physical evidence: ejection marks, firing-pin marks, and latent fingerprint impressions.

While the first two types of physical matching may make it possible to match the shell casing to the firearm, fingerprint impressions left on the shell may match that specific casing to the individual who loaded the gun. Care and handling of the shell casing is of paramount importance. All shell casings should be carefully picked up using only hands or a soft instrument. To avoid destroying any latent fingerprint impressions on the outer surface of the casing, it would be ideal not to handle the exterior in any manner whatsoever. To effect such a recovery, the casing may be picked up by inserting a small piece of wooden dowel into the open end. Admittedly, this may not be practical during most diving operations, so an alternative would be to handle the casing carefully and gently by using the forefinger and thumb, and grasping the casing by the open edge and the opposite end. It may then be carefully dropped into its own container and transported to the surface.

When the casings have been removed from the water, they should be gently and carefully rinsed in fresh water and then air-dried. If they are held together in a magazine, their removal should be done only by a trained and qualified forensic fingerprint examiner. Since drying the bullets may present a problem due to the fact that the magazine may hold water and not allow enough air to circulate, it would be wise to transport the bullets, still in the magazine, directly to an appropriate examiner.

If the recovery is expected to be successful, the investigative dive team may consider having a qualified forensic technician present on the site at the time of the recovery.

All ammunition recovered should be treated as evidence. The rules of evidence follow very plainly. Photographs, while they are a strong asset, are only an addition to the mandatory notes and thorough observations that must be made at the site of the recovery.

Being careful not to stir up the bottom sediment, a public safety diver demonstrates excellent buoyancy control as he retrieves the portion of the barrel from a silt-covered lake bottom. The gun was used to murder four children. It was subsequently tested (ballistics), and because it had been preserved carefully, it was found to positively match the lead slugs found in the body of the victims. The bodies were located by the same public safety dive team approximately 40 miles away.

A sawed-off portion of a rifle was located by public safety divers on the bottom of a lake. Aluminum tent pegs are clearly visible (wrapped in white cord) beside the cut and burnt rifle barrel. The suspect had confessed to throwing both in the water.

RECOVERING DAMAGED OR ALTERED FIREARMS

On occasion a suspect may attempt to destroy or alter the configuration of a firearm prior to disposing of it. He may feel this is added insurance, believing that even if it is located and recovered, the damage he has inflicted will prevent the gun from being used as a viable piece of evidence.

There are many ways a suspect may succeed in destroying the usefulness of a gun as evidence. In particular, two popular methods are burning (allowing the wood to burn and the metal to become red-hot) in an open fire and/or sawing the barrel into smaller pieces. Usually attempts as severe as these, especially when combined with throwing the firearm into water, destroy the characteristics enough to prevent the use of the gun for the purpose of ballistics comparison. This is not always true, however. Burned and cut guns have been successfully recovered from water, and when subject to careful preservation by the investigative dive team and subsequent examination by a forensic firearms examiner, they have been matched to the weapon used in a homicide.

On the Scene: Recover the Pieces

Four youths were sitting on the bank of a river during a warm summer evening. Behind them a very angry and emotionally crippled adult walked up and shot

them before they could escape. Realizing that his initial plan to kidnap at least one of them had failed, he disposed of their bodies by throwing them into the river. The following day, the local public safety dive team was called, and three of the bodies were recovered. The fourth was recovered at a later date.

Several days after committing this crime, the perpetrator disposed of his rifle by cutting the barrel into two pieces and burning them in his fireplace until they were red-hot. After the metallic pieces had cooled, he went to a local lake, swam out a distance of approximately 100 feet (30 m) from shore, and dropped the evidence.

Several weeks later, acting on information received, the public safety dive team searched the lake for the homicide weapon. The pieces were located and properly preserved. They were immediately taken to the forensic crime laboratory, where a slug was forced down the portions of the barrel of the gun using a hydraulic press. Adequate rifling marks were registered on the slugs to confirm that the rifle was indeed the weapon used to murder the four children.

Because the investigative dive team did not give up, they succeeded in recovering both the bodies and the homicide weapon miles (kilometers) and weeks apart.

The suspect was subsequently convicted of four counts of homicide and sentenced to life imprisonment.

MAKING THE DECISION: FINGERPRINTS OR BALLISTICS?

When a firearm is recovered, it should be as certained whether the primary evidence will be in the form of fingerprint evidence or ballistics examination. This must be done prior to the recovery, since firearms preservation techniques will most assuredly obliterate any latent fingerprints. In most cases, the investigating agency will opt for ballistics examination. Latent fingerprints on any submerged item, while possible, are less likely than a positive ballistics comparison.

Properly preserved, a firearm will usually yield a good ballistics comparison. Although fingerprints have been documented on firearms that have been submerged in water for more than two weeks, the chances of lifting fingerprint impressions suitable for identification purposes is, at best, slim. For this reason, firearms are usually prepared for future ballistics examination.

Latent fingerprint examination may, however, be carried out if there is a fingerprint technician on site. As the gun is recovered, the portion of the weapon suitable for examination may be air-dried and examined quickly. This should be considered an option only if it can be done without exposing the firearm to the air for more than a few minutes. Cold freshwater will generally preserve fingerprints on weapons longer than either warm water or saltwater, especially on metal. Latent fingerprints can also be obtained while evidence is still submerged.

When recovering a gun of any type, configuration, or caliber, it should be handled carefully and considered ARMED, LOADED, and READY TO FIRE. Use discretion when handling any unknown firearm. Ideally, it should be grasped only on rough surfaces, such as the wood grips of a revolver or the stock of a long gun. Guns may be raised by passing a cord through the trigger guard.

Recovering Latent Prints from Submergence, Another Alternative: Small Particle Reagent

Many advances in forensic studies have taken place over the years. Some of these include the recovery and removal of latent fingerprints from objects underwater. This technique may be very challenging in many aspects, such as controlling buoyancy, clarity of water (turbidity factors), currents, applying the chemical reagent, photographing and lifting prints.

We can recover latent prints from underwater objects via a procedure incorporating the use of small particle reagent (SPR).

Use of SPR is best when used in the field. During the processing of the crime scene it is always best to collect the evidence and then process it in a laboratory setting. This assists in minimizing contamination during the process. SPR is a chemical agent called molybdenum disulfide (MoS_2). It is used to increase the visibility of and offer further potential for the examination of latent prints. It is like having dusting powder mixed in with a detergent that attaches to latent prints (foreign contaminants such as body oils). SPR can be purchased from a forensic supply house in black or white coloring. It is best to use a contrasting color of SPR to the applied surface. This will allow the latent prints to become more visible to the naked eye. Things to consider when using SPR:

1. Best used when processing from the field from water as well as snow.

2. Best if the latent prints are photographed prior to lifting and making any comparisons.

3. Can only be used on nonporous surfaces.

4. Best if lifted in a controlled laboratory setting.

5. Latent prints may be lifted wet or dry.

6. It's a relatively inexpensive chemical solution.

7. It can be extremely messy, so handle with care.

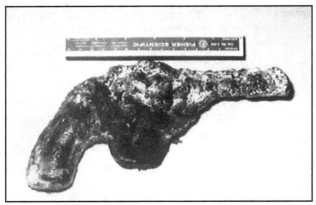

After nearly 30 years submerged, this revolver was recovered by public safety divers. Obviously too deteriorated for successful ballistics comparison, it was still preserved by the dive team using the standard techniques for firearms preservation. This resulted in the gun being restored adequately for a serial number on the handle of the gun to be raised and photographed. After this photograph was taken, the gun was not placed back into a preserving fluid, and the corrosion continued to the point where the serial number could no longer be read. It is fortunate that at least the photographs of the serial number were saved.

PRESERVING THE FIREARM FOR BALLISTICS EXAMINATION

Oxidation (rusting) takes place as a result of oxygen coming into contact with the iron portion of a gun. This process of rusting is accelerated in the presence of water and salts. No matter what the condition of the recovered firearm, it is the duty of the investigative dive team to ensure that the moment the gun is recovered, all further deterioration due to rusting/oxidation ceases.

Obviously, removal from the water and exposure to the oxygen in the atmosphere will accelerate the rusting, yet this is necessary to adequately identify and preserve the firearm. For this reason it must be done methodically and quickly.

The following steps should be followed. Although several of the steps may be considered redundant and unnecessary, the end result will depend on the care taken to preserve the firearm. A complete firearm preservation technique is as follows:

1. Remove the firearm from the water, and carefully immerse it in a bath of clear, fresh water. If the firearm was recovered from polluted or salt water, this bath should be changed several times, while gently moving the gun in its bath to allow the clean water to circulate throughout the mechanism.

2. This step may be completed prior to removing the gun, but it should be done at the first possible opportunity. ENSURE THAT THE FIREARM IS NOT LOADED. It must be handled and/or transported in a safe manner.

3. This step may be considered to be nonessential, but it is very advantageous. Immerse the firearm in pure (anhydrous, containing no water) alcohol or acetone. Care should be taken since these liquids are very flammable. The immersion of a firearm in pure alcohol or acetone will cause the water within the mechanism of the gun to be displaced and the solvent adsorbed, thus preventing any further rusting /oxidation. The process of adsorption is one of loose chemical union between the

alcohol or acetone and the surfaces of the weapon, displacing water molecules. This is a very effective technique to remove all water from the gun prior to immersion in the final preservation fluid.

4. Record the make, model, identifying characteristics, and serial number of the firearm. This may necessitate exposing the gun to the air for short time periods. Do not mark the gun with initials, date, time, etc., unless you are specifically instructed to do so by a superior officer, prosecuting attorney, or qualified technician. If you are instructed to mark the weapon in any manner, record the name of the person who advised you to do so and the specific instructions you were given. The marking of any firearm is a controversial practice.

5. Place the firearm in an airtight container with the final preserving fluid. Perhaps one of the best preserving fluids is diesel fuel or kerosene. This liquid is thin enough to work its way well into the mechanical components of the gun, displacing water and trapped air bubbles while preventing further corrosion of the metal. Both fluids are very flammable, however, and a great deal of caution should be taken when using and transporting them. Another choice of preservation medium would be common motor oil. Either substance will suffice, but experience has shown that kerosene not only preserves but actually restores metal.

6. Seal and mark the container. The container should be sealed using tape, epoxy glue, etc., and marked with the name (initials will usually suffice), date, and time the gun was preserved. The package now becomes evidence that may be transported for further examination. Once properly preserved using this technique, there is no urgency for examination. A firearm that is preserved using this method will not deteriorate any further than it was at the time of recovery.

Although it may be assumed that a successful recovery and preservation will enable the firearm to serve as an exhibit in a subsequent court proceeding, all rules of evidence must be followed.

The notes taken by the investigative diver should include the date, time, and location that any recovery was made, as well as the names of all individuals present.

AIRCRAFT RECOVERY   9

INTRODUCTION

When the subject of aircraft recovery is raised, most public safety dive teams visualize a large commercial airliner that has crashed into deep water. Often the scenario goes even further, with considerations given to the possibility of rescuing passengers trapped inside.

Such is not the case. The reasons for this scenario never occurring are twofold. First, the skin of an aircraft is incredibly lightweight and comparatively delicate. When an aircraft crashes, even into water, it most often becomes a mass of twisted, convoluted, and shredded metal. In addition, due to the relatively light weight of most aircraft bodies, if the hull of the aircraft were to remain intact and enough air remained inside to support life for an extended time period, the aircraft would most assuredly float.

Because of these facts, Hollywood must take a backseat to reality where the recovery of aircraft is concerned. While most public safety dive teams must face the possibility of providing assistance with the recovery of a large commercial aircraft, this possibility is indeed quite rare. Statistics reveal that for every large commercial airliner that crashes into water (or anywhere, for that matter) there are several hundred airplanes with a seating capacity of less than 10 that crash into oceans, lakes, and rivers.

Since most air crash recoveries, rescues, and investigations that the public safety dive team will be called upon for assistance will involve small aircraft, this chapter will be primarily devoted to what might reasonably be expected. If the investigative dive team is educated and prepared to assist in the recovery of a small downed aircraft, then within certain limitations the progression to the handling of a major air disaster will be possible. In short, smaller tasks must be mastered before large ones are attempted. Despite this, however, the safe and efficient recovery of a small aircraft from a body of water is no small task.

AIR CRASH PREPARATION — LIAISON

In most countries, a central agency is appointed as the primary response center and investigative body for all aircraft-related accidents. Most governments, from federal to municipal, have emergency-response capabilities. When an air crash is suspected or reported, the police or fire department is often the primary response center, with the federally-appointed agency arriving quickly and acquiring responsibility for the investigation. In most cases, the local investigative dive team lies quite far down on the notification list, unless that dive team is closely allied with one of the larger agencies.

Prior to any air crash, the fully functioning dive team will form a close liaison with all investigative bodies that may become involved in the search for and subsequent recovery of a downed aircraft. This liaison, once established, should be periodically renewed. If the capabilities (and existence) of the investigative dive team are not known, it will never be called into action.

LOCATING THE CRASH SITE

As in any underwater recovery operation, the first consideration must always be given to locating the object. There are many techniques that have been successfully employed to locate submerged aircraft, but it must always be remembered that the search is not for an aircraft but for its remnants.

Emergency Locator Transmitters

Many aircraft are equipped with emergency locator transmitters (ELTs). The major problems encountered, however, are: (a) on many occasions, they are not activated or turned on by the pilot prior to takeoff; (b) the aircraft is not equipped with a working ELT; and (c) water is not a good conductor of radio waves; hence, the ELT's range may be quite limited, especially in deep water.

Underwater Locating Device

Recognizing the inefficiency of ELT devices, many larger commercial aircraft are now equipped with a water-activated underwater locating device also referred to as a beacon signal. Unlike the ELT that operates on radio waves, the beacon signal utilizes a pulsed acoustic signal to transmit its location. This pulse signal is similar to the signal utilized by both marine mammals (whales, dolphins, etc.) and underwater, wireless voice-communications equipment. Its frequency places it out of human hearing range.

To intercept this signal it is necessary to utilize a specific receiver. These hydrophones are normally of two types: hand-held or shaft-mounted. They are lightweight and easily transported. These devices must be lowered into the water to receive the beacon signal since the signal will travel only a few feet (~1 m) in air.

Locating Devices and Signal Characteristics

There are several beacon signals currently in use. Most commonly, however, they fall within the following parameters.

Operating frequency:	37.5 KHZ
Operating depth:	surface to 20,000 fsw
Pulse length:	not less than 9 milliseconds
Pulse repetition rate:	1 pulse per second (This often varies from unit to unit.)
Operating life:	30 days (variable from unit to unit)
Battery life in beacon:	1 year

Note: There are other underwater locating devices currently used in both the military and civil aviation that for security purposes will not be discussed in this manual.

Snag-Line Search

For the dive team with limited resources, perhaps the most basic search technique is the snag-line search. This search is conducted by using two boats, each trailing a weighted downline. Their downlines are connected by a nonfloating line that will snag the aircraft wreckage. When a snag is registered, the divers may be dispatched to investigate. Should the contact indeed be the aircraft, a permanent marker may be established. This portion (dive) of the search is not without danger, since the divers will be working close to lines that are not only snagged on a bottom obstruction (perhaps the aircraft) but also connected to two boats topside. Diving under these conditions requires training and preparation.

When a snag is made in water too deep to safely dive using conventional scuba equipment, the two downlines may be coupled by using a weighted shackle. Once this is accomplished, they can serve as the actual lifting line. A stage lift may then be accomplished.

A stage lift is a safe technique whereby the lifting device (air bag, etc.) is tied to the line at a point between the surface and the actual object. Once the lift is effected, it may be towed to shallow water, where it may either be lifted or subjected to another stage lift. This is repeated until the aircraft is towed to shallow water.

Training in the use of lifting devices is a mandatory part of any professional public safety dive team's history if recoveries of this type are anticipated. The use of lifting devices should never be considered without first acquiring formal training from a reputable school.

Snag-line searches and stage lifting are not easy techniques. Expert boat handling and good communication between boat operators and divers is necessary. In addition, a snag-line search cannot be conducted over a lake bottom that is covered with tree stumps (as in the case of many manmade lakes and flooded water reservoirs) or shoals and rock outcroppings. A relatively clean bottom is required for the employment of a snag-line search.

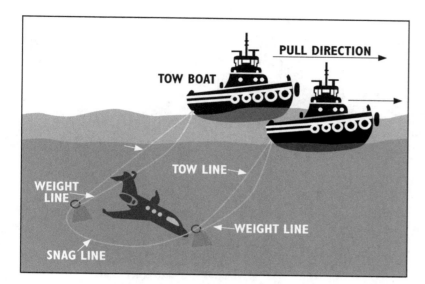

The tow boats are in parallel position traveling at approximately one knot. Note how the weighted lines are attached to the concrete blocks so they may be retrieved up the tow line. Since the tow line is a sinking rope, the portion held between the two weights, that forms the snag line, remains close to the bottom. In practice, the snag line will not form a straight line but tends to form an arc.

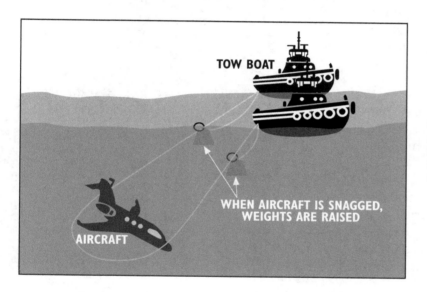

When the wreckage is snagged, the slow-moving tow boats will be drawn together. While keeping tension on the tow line, the weights are retrieved by pulling up the weight lines. The tow/snag lines need to be kept taught.

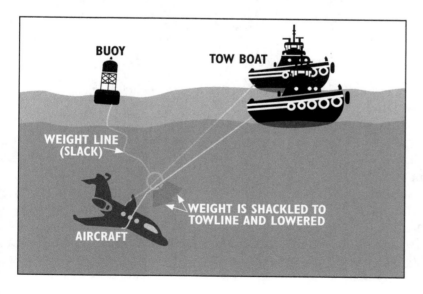

The tow lines are then shackled together enough to allow a noose to form around the wreckage when a heavy weight is lowered down the tow lines. This weight of at least 40 pounds is also tied to a floating line with a buoy. If a log or stump has been snagged, the noose will likely pull free, but in the event it does not, the lines and weights may still be retrieved without the loss of the equipment.

Magnetometers

Instruments such as proton magnetometers are often used to locate shipwrecks, land vehicles, etc. The principle they operate on is one that involves sensing and measuring the change in the earth's magnetic field due to the presence of iron. Since aircraft are made chiefly of aluminum and nonferrous (noniron) alloys, magnetometers are usually of limited or no value in locating air-crash sites.

Sonar

Advanced technology has made sonar a valuable tool. As with any scientific instrument, its use depends on the skill of the operator and the capabilities of the instrument. Sonar schools are conducted primarily for public safety dive teams desiring to acquire capabilities that could be considered the industry standard. The investment of time and money into the use of sonar should be very seriously considered.

The Surface (Fuel) Slick

Perhaps one of the most valuable and least-understood indicators for locating a downed aircraft is the surface fuel slick. When an aircraft crashes into water, in nearly all cases fuel tanks are either ruptured or the fuel in the tank is slowly displaced with water. The sighting of a fuel slick is perhaps one of the easiest ways to determine the location of a submerged aircraft. Since the thickness of the slick may be only a few molecules, it can easily be understood that a very small amount of fuel may provide a large surface slick.

Surface fuel slicks, unless they originate from large quantities of fuel, are usually not highly visible from water level. In most cases, a fuel slick that can be easily seen from an aircraft will be invisible to anyone situated in a boat, even if they are located in the center of the slick itself. The appearance of a surface fuel slick is a result of reflective interference of the sun's rays as they reflect off the thin coating of fuel and off the water surface beneath it. This characteristic appearance, then, is dependent on direct light. It is because of this that fuel slicks are most easily sighted (a) from an aircraft; (b) between the hours of 10 a.m. and 2 p.m.; and (c) on a sunny day. Indeed, a fuel slick that may not be visible on the first day of

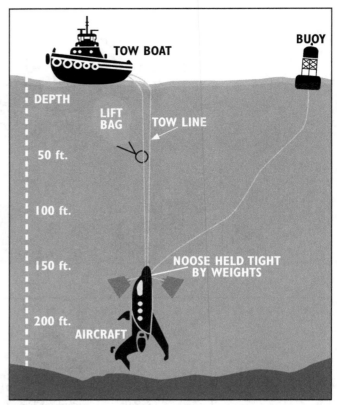

With the weight keeping the noose tight and the irregular shape of the wreckage preventing slippage, the aircraft may now be raised in stages using either a surface winch or an air lift bag. Once lifted a fraction of the distance, it may be towed to shallower water where the stage lifting process may be repeated. The buoy line should be kept free from the tow line (which has now become the lift line) to avoid entanglement.

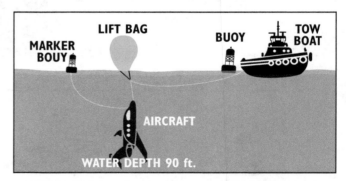

While towing the wreckage, one person should be positioned at the towing point on the boat with an ax or other sharp device that can be used to cut the tow line. This is done if the tow boat is in danger of being sunk by the aircraft. The tow line is between the crash site and the shore. Should the wreckage sink the tow line, it can be cut, and the buoy will then mark the wreckage.

the search if clouds are present may be obvious if the water is again searched (from above) when the sky is clear and the sun is shining directly on the water.

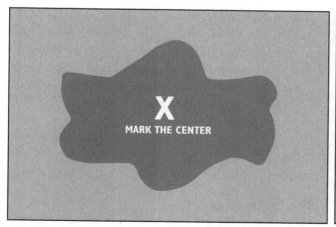

The circular surface slick should be marked at the center. Often, irregular-shaped surface slicks can complicate the location of the center.

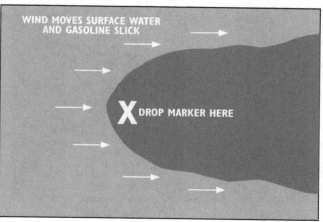

The marker should be set at the prominent upstream point, or head, of the surface slick in a current.

When a search for a submerged aircraft is conducted from the air, it would most appropriately be conducted by first sighting the floating fuel slick then with the use of radio communication, guiding a small boat to the center for placement of a marker buoy. An alternative would be the use of a helicopter; but if an attempt is made to go very low while dropping a marker, the fuel slick may seem to disappear as the helicopter approaches, and agitates, the surface of the water.

THE RECOVERY

Since the recovery of the aircraft and its passengers is usually two different and separate operations, each will be discussed independently.

Recovering the Aircraft

Prior to beginning the actual recovery of a submerged aircraft, its location and the location of any wreckage scattered on the bottom should be recorded permanently. The public safety dive team must be able to return to the exact location days, months, or even years later. Depending on the distance from shore, the depth of the water, and the area over which the debris is scattered, relocating the site may be possible in one of many ways.

A semipermanent marker utilizing a weight that cannot be easily removed by a curious boater may be quickly established and serve to mark the site at least for the duration of the actual recovery operation. A permanent means of returning to the exact site must also be established. This is usually done by carefully plotting the location on a chart. The use of adequately detailed charts may be made more efficient by utilizing several compass sightings, recording bottom contours (sonar), or by the use of GPS.

While the primary site marker may be nothing more than a line tied to the wreckage, it is important to understand that once this wreckage is removed, without adequate marking or recording, return to the site may be quite difficult. This is even more apparent when the wreckage is located a long distance from shore and visual references are less than adequate. For this reason, the site should be marked with at least one (preferably more) markers that are anchored independently of the wreckage. These markers should be located far enough away so as not to interfere with any lifting operations.

Raising the Wreckage

Working closely with the federally-appointed investigative agency is mandatory. In truth, the investigative dive team, no matter how well educated and prepared they are, will be no more than an extension of this agency. The duty of the dive team, therefore, will be to accept direction, make and record observations, and finally, to remove the wreckage. Since the wreckage of most small aircraft is a twisted, contorted, and loosely connected mass of metal, a textbook lift is usually not possible. Most often, after documenting the position of each piece of wreckage, the aircraft will be removed piece by piece. In most cases, the fuselage and/or the tail

Photographed at a depth of 100 feet, the propeller blades of this aircraft were neatly bent backward. No other damage was noted. This clean bending backward would tend to indicate that the blades were not in motion (or at least turning very slowly) at the time of impact with the water.

Unless specifically advised by the air-crash investigator, all airplane parts, pieces, and remnants should be recovered by the investigative dive team. While the propellers may seem very important, they are merely one more piece of evidence. The more evidence that is collected by the dive team, the more accurate a picture will be reconstructed.

section will remain relatively intact, but the forward part of the cabin, the wings, and even the engine, may be missing — lying on the bottom some distance away.

When the actual removal of the aircraft is eminent, the plan must include direction from the investigating agency. Whether air-lift bags, boats/barges with winches, or even the use of a large helicopter is planned, the removal must be done safely and efficiently to the nearest land-accessible shore. Once the wreckage has been taken to shore, it may be inspected closely, photographs taken, instruments removed, etc., in a more leisurely and well-documented manner by the air-crash investigator(s).

Use of Helicopters

When it is decided to utilize a helicopter capable of making the lift, great caution must be taken by the dive team. The operation must be closely planned with the helicopter pilot and all plans approved by the air-crash investigation team. Proper procedures must be followed when working around and under helicopters. In operations such as this, planning and training are not options — they are mandatory.

Use of Air Bags

When the use of air bags (more commonly referred to as lift bags) is planned, the wreckage will have to be raised and then towed to shore. The complexity of raising the wreckage must always be contemplated. Often what appears to be a small portion of fuselage will be found to be connected to many other pieces by cables when raised to the surface. A tail section of an aircraft, once raised to the surface from a depth of 40 feet (12 m) and found to be connected to wreckage still on the bottom, presents a unique problem that is not easily dealt with. Like most problems, it is more safely dealt with by avoidance than by correction. Prior to raising any portion of the wreckage, ensure that it is free and independent from all other wreckage. If this is not done and the resulting "Christmas tree" configuration is left hanging from the lift bag, the dive team is faced with three options.

a) Tow the entire interconnected mass to shore, and in doing so risk further entanglement with the bottom and subsequent loss of evidence.

b) Dispatch a diver beneath the lift bags to free the wreckage. This is indeed a dangerous activity and should be avoided at all costs.

c) Lower the wreckage once again to the bottom, and separate the entanglement prior to beginning a subsequent lift. If this is to be considered, the dangling wreckage should be towed very slowly while being lowered to the bottom. If it is not, the resulting pile of wreckage will be impossible to disengage.

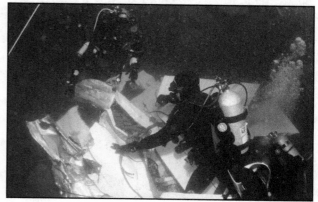

Airplane wrecks are often complex masses of sharp-edged metal pieces. Care must be taken during the lift and tow to securely fasten the mass to lift bags, minimize the probability of diver injury caused by the wreckage, and provide for the quick release of the mass if conditions warrant.

The towing of the wreckage to a predetermined location on shore is a slow process. Most lift bags come with a manufacturer's hitch point, and this, not the wreckage, should be utilized. When towing the wreckage, especially if open-ended lift bags are used, the possibility of loss of buoyancy should be planned for. A quick disconnect/release mechanism must be established on the tow boat. This may be a device like a snap-shackle or may be accomplished merely by passing the tow line over a piece of wood and having a person stand by with a small hatchet or ax.

Bearing in mind the possibility of losing the tow, a secondary marker should be attached to the wreckage. This may be accomplished by simply adding a buoy to the tow line near the boat or towing a marker and line behind the wreckage. Whichever method is used, it should be guaranteed that the length of this secondary marker is greater than the deepest water over which the wreckage is to be towed.

Seizing Evidence and Making Observations

In virtually all circumstances, the seizing of any evidence or exhibits from a submerged aircraft will be done only at the direction of the air-crash investigator. If the aircraft can be brought to the surface intact, it is usually advisable to remove nothing. However, the ultimate decision will rest with the designated air-crash investigator, not the public safety diver.

Aircraft parts that are found on the bottom, separate from the main body of the airplane, should all be retrieved. It is the duty of the chief investigator to determine their value and ultimately retain or discard any items retrieved by the dive team.

The most important responsibility of the dive team (perhaps even more important than the ultimate recovery) is to make accurate and detailed observations. Since the conditions of the wreckage may change during the lift and transport to shore, or articles may be lost from the aircraft, it is important that the public safety diver work closely with the air-crash investigator and/or his team. Under their directions, specific observations may be required and photographs of specific items taken. Underwater photography is a valuable asset for any dive team involved in this type of activity.

In many cases, even before the wreckage is moved, it will be necessary to take photographs of such items as the instrument panel, propellers, and any damage to the body, if indeed there is an intact body remaining. Propeller blades that are neatly bent toward the tail of the plane usually indicate that they were not revolving upon impact. Conversely, blades that are twisted and distorted were likely moving when the aircraft came into contact with the water. The degree of damage to the blades will be assessed carefully by the investigating agency to determine if the aircraft was under power at the time of impact.

Since the force of the impact may freeze certain instruments and controls and yet change others, air-crash investigators tend to look at corroborating evidence, i.e., more than one item, before coming to any conclusion as to the cause of the crash. It is the investigative dive team's responsibility to provide as many clues as possible.

AIR CRASH ONTO ICE

When an aircraft crashes onto a frozen lake, it presents a scenario with its own unique problems and dangers. The public safety dive team that would begin an operation such as this without formal training and experience in under-ice diving would not be performing within the industry standard with regard to safety. Even if trained and experienced in under-ice diving, an air crash through a frozen lake presents a special challenge with its own unique dangers.

The Ice Crash Site

When an aircraft impacts the surface of a frozen lake, its degree of penetration through the ice will be dependent on the weight of the aircraft, the thickness of the ice, the speed at time of impact, and the angle of impact. The angle of impact is most often the determining factor. The following guideline serves as a rough approximation used by air-crash investigators to determine the angle of impact.

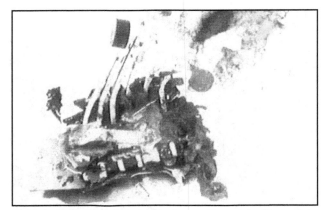

A twin-engine Piper Aztec crashed onto a frozen lake. The angle of collision was estimated at approximately 40 degrees, and speed was believed to be approximately 200 miles per hour. All parts of the aircraft were recovered. The main controls, even though they were draped with the intestines of the pilot, had to be photographed and recovered. This particular evidence supported the estimated speed of the aircraft at the time of impact with the ice.

At an angle of less than 45 degrees, the aircraft may not penetrate the ice canopy. In many cases, wreckage may be strewn over a distance of more than a mile (1600 km) away from the original point of impact.

At an angle of greater than 45 degrees, most of the aircraft will probably penetrate the ice. Depending on the thickness of the ice, there may be little evidence of a crash, other than the original hole along with perhaps fragments from the wings or tail section.

It should be explained that the angle of 45 degrees is not a magic angle where abrupt changes in crash dynamics takes place, but rather a reference to explain a baseline schematic of an air crash.

At an angle of approximately 45 degrees, it is commonly observed that the heavier portion of the aircraft, along with the forward portion of the body, penetrates the ice, while the wings and often portions of the upper body or tail section remain on top. The passengers (who form an integral part of the crash dynamics) are often located respective to the portion of the aircraft they were occupying. On occasion, the top portion of a body may be found several hundred yards/meters (on the ice) from the point of penetration, with the lower portion of the body still contained within the body of the wreckage. The impact with ice often has a shearing effect on the body of the aircraft (and passengers).

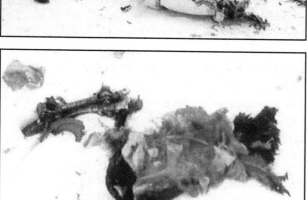

After the crash onto and through the ice of a frozen lake, the aircraft wreckage clearly reveals the direction of travel. This photograph was taken standing directly beside the hole that was left in the ice after the crash. The hole made by the twin-engine aircraft was less than 10 feet in diameter. On top of the ice and underwater, each piece should be carefully retrieved, inspected, and, if possible, photographed prior to removal. The underwater crash site visited by the dive team was even more macabre than what remained on top of the frozen lake.

Very few human remains were located on top of the ice. Most of the passengers were tightly compressed into the remaining tail section of the plane. The recovery operation was complicated by fuel, body parts, and human waste which had collected and been confined by the small hole in the ice.

Perhaps the only advantage of an air crash onto a frozen lake is that its location is well-marked. In most cases, the point of penetration will be surprisingly small, with wreckage strewn in a line, away from the hole, indicating the direction of travel of the aircraft at the time of impact.

The surface of the ice will be strewn with debris from the crash along with (often) fragmented bodies of the passengers.

Hazards

The hazards of ice diving are compounded by the presence of a twisted, confusing, three-dimensional maze of sharp metal that must be negotiated by the diver. Lifelines are easily severed and injuries easily received. In addition to the danger from sharp metal and entanglement, escaping fuel from the aircraft will be held directly beneath the ice canopy and will tend to collect in the hole. This presents a health hazard as well as a fire danger to any personnel involved in this recovery operation. Aircraft fuel is not only a skin irritant but is also very toxic if consumed by the diver. Full face masks and adequate protection are mandatory.

The area directly below the ice canopy should always be searched for bodies. Experience has proven that this location often proves to be a clutter of flotation cushions, passengers' effects, and bodies.

When diving on an air-crash site that is obscured by an ice canopy, all normal hazards exist and are compounded by the cold weather, limited access, and confined fuel slick.

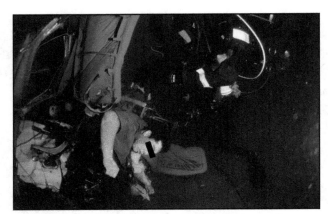
Victims in airplane wrecks often undergo extreme stresses during the event. In this case, the pilot was dismembered at the waist. The lower half of his body was never recovered.

RECOVERING THE OCCUPANTS

Whenever possible, the bodies of the occupants of a downed aircraft should be recovered before the airplane wreckage is removed. There are many reasons for this priority.

a) First, it should be stated that in most instances when an aircraft contacts the water, the forces exerted on the aircraft will also be exerted on the people inside. Obviously, to expect all passengers to be intact and not injured would not be realistic. In truth, diving on a downed aircraft is often a macabre operation and should not necessarily be undertaken by the dive team with little or no experience in body recovery. The bodies of the occupants may be badly disfigured by the crash; in some cases, they will be dismembered.

b) When the bodies of the occupants of the aircraft have been badly damaged or dismembered, they must be removed prior to moving the wreckage. Small portions of the bodies essential to identification must be recovered. Fingers (fingerprints) and upper bodies (teeth) may not be found intact; however, the truth of the situation is simply that these bodies must be recovered. It is not acceptable to leave human remains — any human remains — on the bottom of a lake merely because their recovery was not convenient. Small portions of bodies remaining in the fuselage may fall out during the lift and subsequent tow to shore. This must not be tolerated.

c) If the bodies are removed from the wreckage underwater and carefully bagged prior to raising to the surface, personal effects (jewelry, papers, wallets, etc.) will not be lost. This may be crucial to their ultimate identification.

Note: All human tissue must be removed from the site. The most important body, seen strictly from an investigative viewpoint, will be that of the pilot. A full autopsy on the pilot often supplies important information as the cause of the crash. Unfortunately, because of the twisted and confused nature of the wreckage, a simple seating plan is not often easy to discern. The recovery of all bodies, all parts, and all human tissue found in and around the wreckage is necessary. The presence of a pathologist on site where gross physical damage is encountered may be necessary to document and preserve various tissues that are recovered.

The use of investigative body bags equipped with a fine mesh bottom should be considered so that no evidence is lost, yet the bodies (and all associated evidence) may be removed from the water and into a boat easily.

AIRCRAFT RECOVERY— HAZARDS

The recovery of submerged aircraft carries with it its own unique procedures and responsibilities. Likewise, the public safety dive team should recognize that there are hazards connected to this activity that are not found in other recovery missions.

The number and variety of potential hazards that may be found during any recovery of a submerged aircraft are unlimited. The following list of hazards, however, should serve to alert the dive team to specific dangers and to raise the awareness of the team to a level of safety that can only be referred to as professional.

While this is not an all-inclusive list, many or all of these hazards could easily be encountered during any aircraft-recovery operation.

Sharp Metal

When the thin metal skin of an aircraft is exposed to crash forces, it is often stretched and torn,

Control cables present in most smaller aircraft are often difficult to see, especially in murky waters. The ever-present potential for diver entrapment must always be foremost in the minds of any investigative dive team involved with air-crash recoveries. These may involve the recovery of a wreckage of metal that has been twisted, torn, stretched, and strewn over a wide area — often connected by cables and loose debris.

The hazard presented by sharp metal cannot be over-emphasized. When an airplane crashed, this aluminum skin is often stretched and torn like toffee. It then presents a razor-sharp edge that may cut lifelines, suits, and divers.

leaving razor-sharp edges. Diving equipment, suits, lifelines, and divers may be easily cut if proper precautions are not taken. All edges should be considered sharp and appropriate precautions taken, both during the search of the wreckage and the subsequent lift.

Fuel

It should be easy to remember that if the crash site was located by a visible fuel slick, then divers will be required to swim in and descend through this slick to reach the wreckage. Unfortunately, each year dozens of public safety divers ingest quantities of aircraft fuel and experience painful skin and eye irritation through exposure to fuel escaping from a submerged aircraft.

Even in cases where there is no apparent fuel layer floating on the surface, once the aircraft has been disturbed, especially during the lifting procedure, large quantities of fuel may be released. This may present not only a health hazard to any divers still in the water, but also in cases where large quantities of fuel escape, the potential of a fire hazard must not be ignored.

Cargo

Before diving on any aircraft wreckage, an accurate manifest of cargo should be obtained. Aircraft are vehicles of transport, and as such carry substances that may result in injury or death of the diver if he is not adequately prepared and equipped to safely handle such a situation.

Cargo such as pesticides (crop dusters) and herbicides present deadly combinations. Other cargo such as chemicals (i.e., potassium cyanide, often transported into mining camps) is deadly when dissolved in water.

In addition to legally transported cargo, small aircraft are often used to transport explosives to remote locations. This is against most federal regulations. The fact that an aircraft may contain explosives may not be readily available to the dive team. The containers containing explosives, however, are usually clearly marked. This, however, does little to inform the public safety diver who is diving in reduced or zero visibility.

Aircraft Equipment

While not truly considered cargo, most larger aircraft contain high-pressure oxygen tanks that could very likely become damaged upon impact. Most divers are well aware of the shape (and feel) of high-pressure cylinders. When oxygen cylinders are located on a commercial aircraft for the purpose of supplying oxygen in the event of cabin depressurization, the cylinders are usually turned on (their valves open), and the lines carrying the oxygen are pressurized.

Fire extinguishers aboard aircraft may also provide hazards not normally encountered by the public safety dive team. In many larger aircraft, each engine is equipped with its own fire-extinguishing mechanism. Fire extinguishers usually present a combination of cylinders capable of providing pressure and/or chemicals quickly. Either situation is dangerous for the diver to encounter.

Cables

Control cables present in most smaller aircraft are often nearly invisible in limited-visibility conditions. These cables are very resilient and virtually impossible to cut should entanglement occur. Any movement in and around submerged aircraft should be done slowly, carefully, and with appropriate extrication procedures in place should entanglement occur.

Tires

Large commercial aircraft employ high-pressure pneumatic tires, which if ruptured would most certainly kill any individual nearby. In large commercial aircraft crashes on land, the tires are often deflated at a distance, using a high-powered rifle. The public safety diver does not have this option. High-pressure pneumatic (air-filled) tires should never be handled.

Hydraulic Lines and Systems

Hydraulic lines and systems on larger aircraft have the potential to severely injure the diver. Hydraulic lines often contain pressures well in excess of that found in the average scuba cylinder. If ruptured, both the (potential) explosive force as well as the hydraulic fluid itself represent serious hazards for any personnel involved in the recovery.

Exit Signs

Large aircraft often utilize cautionary exit signs (for emergency exits). In many cases, these signs contain a radioactive material that should never be allowed to come in contact with the skin or any portion of the body. These phosphorescent signs become their own warning of danger: If it glows, do not handle it.

Ejection Seats and Munitions

On occasion, public safety dive teams have been the first respondents after a military air crash. Military aircraft carrying live ordnance should be avoided. This type of recovery carries with it a special danger that can be dealt with only if the diver possesses the specialized knowledge and training to safely deactivate and remove military ordnance.

Ejection seats, if activated by accident, will likely prove fatal to the diver who is attempting to remove the pilot from the aircraft. Without proper training, this type of recovery operation should be avoided.

VEHICLE RECOVERY

10

THE IMPORTANCE OF VEHICLE RECOVERY/INVESTIGATION

Each day several thousand vehicles are reported stolen in North America. Understanding why they are stolen and who steals them becomes the first step in the efficient recovery of the automobile by the investigative public safety dive team.

There are three major categories of incident in which public safety dive teams may be called upon to recover vehicles that have been submerged in water. They are the auto theft, the crime vehicle, and the motor vehicle accident.

Each type of recovery places specific responsibilities on the public safety diver. These responsibilities and tasks may be effectively carried out if certain facts are first understood. These facts include how and why vehicles are stolen, and how and why they are disposed of in water. Regardless of who, how, or why the vehicle was stolen, or even how it subsequently came to rest in the water after a motor vehicle accident, the vehicle and its contents must always be considered as an exhibit — an item that if inspected with care and properly handled will tell a story.

RECOVERY OF THE STOLEN VEHICLE

Thefts of motor vehicles will usually fall into one of three categories: the joy ride, the professional theft, or fraud. Each type of theft carries its own unique evidence indicating which type of theft was involved and the identity of the perpetrator.

The Joy Ride

Perhaps the most common of all automobile thefts is perpetrated by youths who steal a car simply for the purpose of joy-riding. Usually this entails a brief period of local driving, often coupled with drinking alcohol and culminating in the disposal of the vehicle in the easiest and most opportune manner.

The Theft

The target vehicle in these cases is usually the easiest vehicle to steal. This crime of opportunity usually targets vehicles that have had their key left in the ignition. On occasion, windows are broken, and the ignition mechanism is forced using a tool. The most common tool utilized by youths is a heavy, long-handled, slotted screw driver that is bent at 90 degrees. Once inserted in the ignition keyway, the tool is then twisted, allowing the ignition cylinder to turn. The ignition keyway is broken, but the engine will start and the steering-column locking mechanism is disengaged. Often the entire ignition mechanism is removed, wires are stripped, and the vehicle is hot-wired. When an extended tour is contemplated, the perpetrator(s) may secure the wires with tape and tuck the entire mechanism back into place, complete with a dummy key.

Entry into locked vehicles by untrained culprits is usually through unlocked doors or by breaking a side window. Since the breaking force in these cases is always exerted inward, a quantity of glass will usually be found inside the vehicle, on the floor, and the seat, even after recovery from the water.

Another technique becoming popular with would-be car thieves is the use of porcelain chips. Pieces of porcelain that have been removed from a spark plug (usually by using a hammer) are thrown forcefully against the side window of a car. When these hard, sharp chips are thrown at a side window, the window instantly crumbles inward. Where the recovered vehicle displays a broken side window, the interior should also be searched for tell-tale white porcelain chips. This technique for breaking windows has become very common because of the speed and relative silence it allows.

The uneducated or amateur car thief rarely resorts to the use of lock picks, Slim Jims or other sophisticated devices.

The Vehicle

The casual thief prefers a high-powered sports car but will settle for any vehicle of opportunity. Since most joy rides involve vehicles that have been left running, or at least left with the keys in the ignition, this type of car theft often occurs in business districts during daylight hours. The location and type of vehicle takes a lower priority than the ease of the theft.

The Disposal Site

When the amateur car thief eventually decides to dispose of the vehicle, it is usually left by the road or parked at random. Similar to the ease of the theft, the ease of disposal remains a constant. The casual or amateur car thief is an unsophisticated, lazy individual who usually takes an easy route.

On occasions where the vehicle is disposed of in water, it is usually as a result of several youths acting in concert and thinking, "I wonder what would happen if...." In most cases, the site is a boat launch ramp at a local lake or marina. The ease of entry into the water here is the key to understanding "who" disposed of this vehicle. Most boat launch ramps have a gentle slope into the water that requires only a car and a transmission left in neutral. Gravity supplies the forward motion. These cars are usually located very close to shore, since they entered the water at a slow speed.

The Search for Evidence

In the majority of cases, the vehicle recovered will be in neutral, the key will be in the ignition, and the ignition may or may not be in the on position. If the vehicle was disposed of after dark, the headlights will usually be turned off, since the thief would not want to draw attention to his actions.

When it is discovered that the ignition has been removed and the wires taped together, fingerprints may be left on the glue side of the tape. This tape should be removed and inspected only by an investigator trained to do so. Once the vehicle has been removed from the water and sufficiently dried, other sites may be examined for fingerprints. These sites include the rearview mirror, windows, steering wheel, or any clean smooth surface, including the knob placed on the seat adjustment lever.

Vehicles stolen by youths are often utilized as a mobile party site. If such was the case, beer tins, bottles, and other containers that held alcoholic beverages found inside the car should be carefully removed, dried, and examined for latent fingerprint impressions. Fingerprint impressions have been found on glass in cases where submergence exceeded two months.

Even before the vehicle is removed from the water, a cursory inspection should be conducted for body damage. Many stolen vehicles are involved in hit-and-run accidents. If damage is noted to a specific area, further damage to that area should be avoided at all costs during the removal of the vehicle. After removal from the water, measurements, photographs, and paint samples may be removed for further investigation and possible matching to a local hit-and-run victim vehicle.

Car radios, stereo equipment, hubcaps, and other valuable, easily-removed items may be missing from vehicles recovered after a brief joy ride. These items are usually characterized by their ease of removal and ease of sale. In cases where car stereos, etc., have been forcibly removed, pry marks may leave impressions that eventually could be matched with a tool found in the possession of a suspect arrested at a later date.

The site where the casual or amateur car thief usually chooses to dispose of a car in the water, besides being a convenient one, often corresponds to a location that local youths and youth gangs are known to frequent after dark. The entire theft and disposition of the vehicle then becomes a game of show and tell, perpetrated for perhaps no other reason than peer acceptance.

THE PROFESSIONAL THEFT

There is an incredible black market specifically for parts of high-priced vehicles. Most luxury cars that are stolen by the professional for this market are reduced to parts within 24 hours. The remaining body (often only a chassis) is disposed of quickly in as convenient, remote, and secure a place as possible. To dispose of stolen parts, the thief must engage the automobile industry at some level, whether at a body shop, parts recycler, new and used car dealer; somewhere the stolen part must re-enter the marketplace. Most states have strict laws regarding licensed vehicle dealers to prevent or at least track these types of crimes.

The Theft

Professionals often use tools such as Slim Jims to enter a vehicle. These long flexible strips of metal are fashioned so as to easily slide down between the driver's window and the rubber weather strip. Once in place, they are fashioned such that the linkage leading to the external lock can be snagged and manipulated to unlock the door. The use of a Slim Jim is not a simple task. Several models are commercially available, and various makes and models of vehicles require the use of specific tools. The use of such a tool requires a trained individual who understands how to enter a vehicle quickly and quietly.

Once inside the vehicle, the professional often removes the entire locking mechanism and replaces it with his own cylinder, complete with a working key. Once installed, it is not detectable during a routine stop by the police. The mechanism works, the key fits, and all seems normal. In addition, the professional will usually be reasonably well-dressed and carry no open liquor during the theft or subsequent transport of the vehicle to its ultimate destination. He does not want to arouse any suspicion.

Prime targets for the professional car thief are underground car parking lots, apartments, and business establishments. The professional usually requires either the privacy of seclusion or the privacy of a very busy parking lot. Since a well-practiced professional can enter and start most vehicles in under one minute, he requires only privacy from the owner of the vehicle. The thief's exit in this vehicle is usually quiet, and his driving habits are such that they do not attract any undue attention.

Usually, there will be a customer waiting for the vehicle, and it will be delivered within the hour. This customer is usually referred to as a chop shop, a location where the stolen vehicle is quickly disassembled into spare parts that are (a) easily transported, (b) not identifiable, and (c) in high demand. Thus, doors, fenders, bumpers, and on occasion even engines and transmissions, are removed and immediately transported to a local buyer.

The Vehicle

Most vehicles stolen by professionals who are employed in this trade fall into the luxury class. The average vehicle owned by a middle-income family would not bring the profit necessary to support this professional network. While there are exceptions to this rule, in most instances luxury vehicles less than three years old are prime targets for the professional.

Another exception that should be mentioned is the luxury vehicle that is stolen, then through a complicated set of maneuvers is retitled in a distant state/province or even country. This usually is the action of a lone professional who is able to disguise or change the vehicle identification number (VIN) upon arrival at his new location. The vehicle is often given a new "clean" VIN and title from another crashed vehicle of the same make and model, and the vehicle is usually sold quickly for cash at a greatly reduced price. In most cases the naïve new owner discovers the problem during a future sale or registration of the vehicle.

In some cases the thieves may make their own VIN plates, however gaining the rosette rivets to install them can foil their efforts. Currently some vehicle manufacturers are discontinuing the use of rosette rivets in favor of electronically made VIN plates. Law enforcement agencies have access to electronic plug-in devices that can read the true VIN from the electronic control unit on a suspect vehicle, and this helps solve the vehicle identification problem.

The Disposal Site

Professional auto thieves must find a location for all vehicle parts/remnants that remain after the planned stripping of the stolen car. Often water presents them with the ideal site. The site often chosen is one where the culprit feels the evidence will not be found by accident. Local marinas and fishing rivers do not present such a site. In many cases the body of water that is chosen is deep (such as would be experienced at the base of a cliff overlooking a lake or quarry) or murky and fast-moving (such as would be found in a large river). The site is usually carefully chosen so that the wreckage is not accidentally found by swimmers, boaters, or even recreational divers. These sites are usually either remote or are utilized during hours of least traffic (3 a.m. to 5 a.m.).

The Search for Evidence

When searching the site of what may be suspected of being a professional drop-off zone, it should be remembered that this specific area may contain the remnants of several vehicles. Indeed, locations containing the remnants of more than 30 vehicles have been the site of recovery operations that have lasted for weeks.

Rarely will vehicles be found intact. When an assortment of vehicles is located in one confined area, and these vehicles are relatively intact, suspicion should be directed away from the professional thief and concentrated on a youth who lives in the area.

All vehicles are marked with an assortment of serial numbers. Engines, transmissions, radios, and other accessories often carry independent serial numbers that may be traced back to the manufacturer and then forwarded to the most recent owner. Vehicle identification numbers, which normally appear just below the windshield on the driver's side of the vehicle, are often missing when a professional thief is the culprit. This VIN plate invariably finds its way to another vehicle/owner in due course.

Despite the customary removal of the VIN, all vehicles are equipped with at least one confidential number ("con number" or "secret VIN"). The location of the secret VIN is determined by the year, make, and model of the vehicle. For any specific vehicle, the location of the secret VIN may be obtained directly from the manufacturer. While this serial number is referred to as a secret VIN, the location of such numbers is not at all a secret and may easily be located by inquiries through manufacturers, retail car agencies, national auto theft bureaus, the National Insurance Crime Bureau (NICB), etc. The secret VIN, however, is not as easily removed as the window-mounted VIN plates and is not as easily accessible.

Papers placed in the vehicle in certain areas are called "tell slips" and have identifying information that will prove helpful in many cases. Tell slips are often found under coverings on the seats and elsewhere.

Vehicles and vehicle parts removed from a dumping ground used by a professional auto thief or auto-theft ring should be carefully handled and examined for tool impressions that could be ultimately matched to the shop that disassembled the vehicle.

The public safety dive team that must recover automobile remnants from what appears to be a professional dumping ground should recover all items.

On the Scene: Watered-Down Prices

A small public safety dive team was given the task of removing several car chassis from a limestone quarry.

Early into the recovery it was noticed that this site contained the remnants of perhaps as many as 20 vehicles, all late-model luxury cars. Because of the various degrees of algal growth on the parts and chassis, it was determined that this site was currently in use by the perpetrator. The newly formed dive team, in an effort to impress their sponsoring police agency with their ability, recovered everything, including a socket wrench in excellent condition. Upon further inspection of this wrench, it was noted to have a driver's license number engraved on the shank. Inquiries led to the subsequent location of the owner, who had opened a business in a neighboring county: an auto-wrecking business. Warrants were obtained, the premises searched, and his tool box was found to be missing his socket wrench — and nothing else.

The business owner was convicted of possession of stolen property, and more than $200,000 retail value in parts was seized. He had been convicted of similar offenses in the past and had just been released from prison the previous year. Strangely enough, the advertisement in the local newspaper that had run for several weeks prior to the arrest read: "BILL'S IMPORT CAR PARTS & ACCESSORIES. WATERED-DOWN PRICES. YOU ORDER IT...WE'LL FIND IT."

The Fraud

In cases involving fraud, the theft is usually perpetrated by the car owner or an associate. The reason behind the theft is usually an attempt to collect insurance money. Because of this, the owner is most often a person who has a problem that requires money as a solution. Insurance policies in such cases usually have a very small deductible clause and offer full coverage for theft.

The Theft (Crime)

Until recently, most vehicles that were disposed of for the insurance money were driven to a remote location and set afire. Since insurance companies have improved their investigative capabilities, and more police departments automatically investigate a car fire as a suspected fraud, this method of disposal has become less common. Since youths dispose of more cars by driving them into the water than setting them ablaze, more and more people are choosing to mimic the youths and simply drive their car into a nearby lake, river, quarry, or ocean. The theft is then reported to the local police department, and upon recovery of the vehicle, insurance is paid quickly.

The Vehicle

Typically, the vehicle that is disposed of to claim insurance coverage is usually a late-model luxury vehicle. Its actual time of theft is often reported in a vague manner ("sometime during the night" or "over the weekend — I'm not sure, I wasn't around and didn't check"). In most cases, there will still be a sizable outstanding loan covering the amount owed on the vehicle, and the owner may be experiencing financial difficulties. The solution is always the same. Rather than have the vehicle repossessed and lose all his money, the owner creates a "stolen" vehicle, the insurance is paid, and the remaining portion of the loan is paid off with some cash remaining for the owner.

Another indication of an insurance fraud is often a large repair bill coupled with staggering monthly payments. Often when an owner of a luxury vehicle sees his purchase as a "lemon" costing more money than anticipated just to keep the vehicle operating, disposal becomes a viable option.

A small piece of wood was used to force this accelerator pedal down. In such cases, the vehicle's automatic transmission lever is usually found in the drive position and the driver's window is open. With the vehicle pointed toward the water, the perpetrator reached through the window with a stick, hit the gear shift lever to put the car into drive, and sent the car accelerating into a body of water where it was later recovered.

The Disposal Site

The disposal site is frequently similar to the type used in the joy ride. Since many insurance policies will pay the owner only after recovery of the vehicle (or a lengthy period of time), the goal of the perpetrator is to simulate a theft, destruction of the vehicle, and a speedy recovery. These conditions limit the choice of a disposal site. Ideally, a shallow body of water is chosen so that the vehicle may be discovered by independent/innocent parties shortly after it is submerged. Boat launch ramps are ideal for this purpose, and many vehicles resting in shallow water have been discovered a few inches (centimeters) below the surface as pleasure boats scraped their propellers across the roof. Other sites include shallow ditches and clear quarries. When salt water is an option, it is usually chosen over fresh water disposal in an attempt to render the vehicle totally incapable of being repaired and returned to its owner in lieu of a monetary payout.

The Evidence

As with other types of crimes involving the ultimate disposal of vehicles in water, the fraud vehicle leaves behind its own unique trademark. Unfortunately, the greater portion of the clues that lead ultimately to a suspicion of fraud involve good investigational techniques outside of the domain of the volunteer public safety dive team. Despite this, there are certain clues that could easily be lost during the recovery but that should not go unnoticed.

When a vehicle is disposed of by its owner, it will be stripped of any material or items that carry an emotional value. In most cases there will be no family photographs, no address books, and no personal papers of any continuing value. Similarly, favored dashboard ornaments will be removed prior to the theft, and if the vehicle is equipped with a CD player, most if not all of the CDs will be missing. If the owner is a married individual with a family, children's favored toys and clothing will usually be removed as will infant-restraining car seats. In brief, the vehicle will be devoid of virtually any item that would carry significant emotional attachment to the owner (but not necessarily other family members). If the marriage is in the process of breaking up, there may be items of great emotional value belonging to the wife — these items almost assuredly would be easily susceptible to water damage. Their presence in the vehicle displays what is commonly referred to as the coward's revenge.

In a fraud vehicle that is recovered from the water, usually the evidence is simple. The driver's window is down, the ignition key is in the on position, the gear selector is in drive, and some object is utilized to depress the accelerator pedal. In many cases, the accelerator is depressed simply by placing a rock, concrete building block, or some other heavy object on it. In many instances, a small piece of wood may be wedged between the pedal and the dash or the seat to depress the pedal. These items are usually discovered in place by the divers who first make their observations but become easily dislodged during the recovery phase of the operation.

Of late, a rather unique approach has been utilized whereby an inflatable object (a child's inflatable toy is a common choice) is used as a wedge to depress the accelerator. Once the vehicle has settled on the bottom, even in 6-10 feet (2-3 m) of water, the pressure exerted by the water causes the inflatable toy to shrink in size (volume) and float freely away from the accelerator. Many vehicles recovered in these cases have been noticed to have their headlights left on, presumably to draw attention to the submerged vehicle by passersby. This would assist with a speedy recovery and ultimate insurance payoff.

In cases where the vehicle was disposed of for insurance reasons, there will be no attempt to remove the license plates, VIN, tag, or even registration papers. Indeed,

up-to-date papers are often found in the glove box or strapped to the visors sealed in watertight plastic bags. Every (reasonable) effort is usually made by the owner to assist the police in locating the "victim" whose car has been "stolen." Indeed, in many instances where owners have moved many months prior, their registration papers are updated with the department of motor vehicles just prior to the alleged theft.

In rare instances, the owner of the vehicle will make an attempt at vandalizing the vehicle prior to disposing of it in the water. Such attempts are usually aimed at the glass or the body of the car, but this is rare since it would only serve to attract attention to his already illegal activity.

RECOVERING THE CRIME VEHICLE

Vehicles involved in crimes such as armed robbery, kidnapping, rape, or other serious offenses are often disposed of by submergence in water in an attempt to either destroy or hide evidence. This is particularly true when the vehicle is either stolen or belongs to the victim. Unfortunately, victims of crime are also sometimes disposed of with the vehicle. For this reason, a body found in a submerged vehicle should never merely be considered to be the driver or passenger in a simple motor vehicle accident. A rule of thumb seems to exist whereby the degree of ingenuity displayed in disposing of evidence is in direct proportion to the seriousness of the crime. It then follows that when a crime vehicle is disposed of by submergence in water, it may not only be difficult to find but also may be or contain evidence of a very serious crime.

Just as all accidental deaths must be investigated as homicides until the latter is ruled out, all vehicles should be assessed as crime vehicles until their origin is known. If this procedure is followed, evidence should not be lost or overlooked. For an extensive list of types of evidence that may be collected from a crime vehicle, refer to Collection and Preservation of Evidence, and Common Crime Scenes.

When dealing with the recovery of a crime vehicle (or any vehicle), it should be remembered that as the vehicle is raised to the surface and/or towed to shore, considerable evidence may be lost. It is necessary that the fully-functioning dive team be prepared to carefully enter the vehicle and make observations,

A vehicle used in a homicide has been towed from the water. Prior to its removal, the investigative dive team removed all evidence possible before securing the doors and windows shut. The vehicle was removed from the water very slowly and allowed to drain over a period of 30 minutes.

take photographs, and seize exhibits whenever possible. If conditions such as strong current and poor visibility do not allow this, the very minimum protection afforded by raising all windows and closing all doors should be attempted. The actual removal of the vehicle from the water should be done very slowly with great care, allowing the water to flow out of the vehicle in a controlled manner. It is not unreasonable to take as much as a half hour from the time the vehicle first surfaces on the end of a tow cable until it is fully removed from the water.

The efficient recovery of a crime vehicle is an exercise in observation and exhibit recognition/preservation.

MOTOR VEHICLE ACCIDENT

The primary responsibility of the public safety dive team involved with the recovery of a vehicle involved in a motor vehicle accident is twofold: (1) recovering the victim(s); and (2) recovering the vehicle.

Recovering the Victim

When a vehicle leaves the road, wharf, boat, or bridge, its occupants are often overcome with panic, injuries, or cold water. Dive teams are often called upon to recover the bodies of these victims.

Prior to discussing the mechanics of this type of body retrieval, an important fact to understand is that when a vehicle enters the water (even from a height of 10-15 feet/3-4.6 m), rarely does a vehicle sink in less than one minute. Indeed, even with the windows rolled down, many mid-sized vehicles will

remain afloat for several minutes. In most cases, an immediate attempt is made by the occupants to open the doors. Against the pressure exerted by the water outside, this becomes nearly impossible. Rational thought would normally guide them to climb out the open window(s), but rational thought is not a common quality in people who have found themselves trapped inside a floating automobile.

As the water level rises to the window level there may be an attempt to close the windows. If the windows are raised, the ultimate sinking of the car is delayed temporarily, but if the windows remain open, as the water level rises over the level of the windowsill the car usually noses forward, and water rushes inside in a rather forceful/violent manner. As the vehicle begins its descent, a remaining air pocket may be temporarily established inside the vehicle passenger compartment. An attempt may be made by the occupants to migrate to this air pocket, but it quickly escapes through the backseat partition and out through the trunk. It is for this reason that the trunks of vehicles that have been found submerged in more than 30 feet (9 m) of water are often found with their trunk lids open. The escaping air through the trunk cavity has temporarily deformed the trunk lid and forced it open. Usually there is no obvious damage to the trunk lid, only to the latch mechanism.

Once this process is understood, it can easily be explained why drivers and passengers of vehicles are often located in the rear portion of the vehicle.

Consideration of whether to remove victims separately from the vehicle removal should be reviewed by the public safety dive team. If a vehicle is secure and can be made to remain so, and careful removal of the vehicle is possible, it may be wise to follow that path. If on the other hand the vehicle has been breached in some way, as in doors opened, windows down, etc., then collection of victims and contents would be appropriate underwater. Prior to removing any occupants of a submerged vehicle, their positions should be noted and photographed where possible. At the very least, comprehensive notes should be taken. Once this is done, their removal should be executed carefully, placing each individual in his own body bag prior to transport to the surface. Body bags that are equipped with a mesh bottom are ideal for this purpose, since small personal effects are not lost during transport or subsequent removal from the water.

Vehicle registration papers should be carefully removed and placed in rigid containers for protection along with any other articles that may have a bearing on the incident. A seemingly discarded envelope on the floor of the vehicle could contain a suicide note or a recent repair bill that could assist in explaining the accident. Not losing critical items is the goal of the underwater removal of victim(s) and other evidence prior to vehicle removal. Consider using lift bags to minimize damage and assist in controlling the vehicle recovery. All rules of observation and conduct involving any body recovery should be followed.

The Vehicle Recovery
A great deal of care must be exercised when removing a vehicle that has been involved in an accident. The vehicle itself is a piece of evidence, and a mechanical inspection may provide distinct clues as to the cause of the accident. For this reason alone, further damage to the vehicle body during the removal process may hinder calculations that would be performed by an accident analyst. In addition, damage to the undercarriage could mask or even obliterate hard evidence that could otherwise have been used to explain the accident.

If lift bags are to be utilized to raise the vehicle, they must be attached to points of strength inherent in the structure of the vehicle. The wheels afford excellent lift points, and a four-point lift is always advisable. An easier method of merely passing a chain or strap under the roof and through the passenger compartment is often used, but this should be avoided, since the forces exerted by the lift could deform the body and frame of the vehicle.

Whether a vehicle is lifted to the surface, then towed to shore, or merely towed into shore using a tow-truck cable, it must be done in such a manner as to avoid any further damage to the vehicle. On a flat bottom, the steering wheel may be secured and the vehicle towed directly to shore. Where debris precludes this simple recovery, the vehicle may be lifted over bottom obstructions, but

As a vehicle leaves the surface, the weight of the engine forces the front end down. Air inside the vehicle migrates quickly to the rear, often exiting through the trunk and forcing the trunk lid open. Although the trunk lid is not always opened in this fashion, care should be taken before coming to the conclusion that it was forced open prior to the vehicle entering the water.

the main principle remains unchanged:

When recovering evidence — first do no harm.

Consulting the Accident Analyst
Most police departments either employ or have access to trained motor vehicle accident analysts (traffic accident reconstruction specialists). Given specific information, investigators skilled in this area of expertise are able to reconstruct the details of an accident, using as a starting point the final resting place of all vehicles involved and the degree and type of damage to the vehicles.

The role of the investigative dive team becomes incredibly important if the accident analyst is ultimately to be consulted. In cases where an analyst is to be called in to assist in the investigation, he should (when possible) be consulted prior to moving/recovering the submerged vehicle. Specific requests may be made as to underwater photographs, measurements, and observations that may be lost during the recovery phase of the operation.

Dive teams that work closely with trained accident reconstruction specialists are able to effect the recovery of a vehicle and preserve all physical evidence.

It is indeed unfortunate that investigative dive teams often ignore the services of traffic accident reconstruction specialists merely because they did not know they existed. Indeed, the same excuse is often used by police departments when the local public safety dive team is not called upon simply because the police department did not know of its existence.

It is the responsibility of the fully functioning public safety dive team not only to advertise its presence and expertise to all concerned agencies, but also to maintain a list of other support services available. While the investigative dive team may not be responsible for contracting the services of a qualified traffic accident reconstruction specialist, the team is certainly able to make recommendations to the police department that has contracted for its services in any vehicle recovery.

As in all recoveries, the role of the investigative public safety dive team is to recover all evidence in its original condition..

RECOVERING HYBRID AND ELECTRIC VEHICLES

With advancement in the diving industry also comes advancement of hybrid vehicles. Hybrid vehicles require special attention as their systems operate off fuel, battery, and different methods of power generation. Because of the growth of this segment of technology, dive teams should take a nationally recognized class on how to best approach these vehicles.

Most cars are equipped with a fail-safe computer module that is designed to kill the power. Response protocols suggest you avoid contact with high voltage areas, follow departmental practices and wait until the vehicle is on land to disable power. Other potential areas for exposure are damaged batteries and fuel leakage issues. See the manufacturer for specific recommendations for submerged vehicle practices.

AUTOMOBILE RECOVERY CHECKLIST

Investigator's name _____

Department _____

Date _____ Time of recovery _____

Names of those present during recovery (witnesses names for court purposes should be recorded)

_____

_____

_____

(Use extra paper if necessary.)

1. Is the car's transmission in gear?   ❏ YES   ❏ NO

2. Is anything blocking the accelerator?   ❏ YES   ❏ NO

3. Is anything in the car that could serve   ❏ YES   ❏ NO
   as an effective block for the accelerator
   (block of wood, stick, inflatable toy)?
   Describe (given location).

_____

_____

_____

4. Is the radio on?   ❏ YES   ❏ NO

5. Is it tuned to a station commonly   ❏ YES   ❏ NO
   listened to by the car owner?

6. Any cigarettes or liquor in the vehicle?   ❏ YES   ❏ NO
   List brand name and location found.

_____

_____

_____

7. Any "emotionally valuable" items left in the car (children's toys, ornaments, etc.)? ❑ YES ❑ NO

8. Any other valuables in the car or trunk (mechanic's tools, camera, business papers)? If yes, list items (give location where item was found). ❑ YES ❑ NO

_____

_____

_____

(Use extra paper if necessary.)

9. Headlights. ❑ ON ❑ OFF

   Interior lights ❑ ON ❑ OFF

10. Is point #9 consistent with the time the vehicle went into the water?

    ❑ U/K ❑ YES ❑ NO

11. Is point #9 consistent with the time the vehicle went missing?

    ❑ U/K ❑ YES ❑ NO

12. Were the windows(s) up or down when the vehicle was recovered?

    Driver's window ❑ UP ❑ DOWN

    Passenger's window ❑ UP ❑ DOWN

13. Is observation #12 normal for the time of the year or weather?

    ❑ U/K ❑ YES ❑ NO

14. Ignition switch (key) position?

    ❑ ON ❑ OFF ❑ AUXILIARY ❑ NO KEY

15. Does the position of the driver's seat compare with the size of the car owner?

    ❑ U/K ❑ YES ❑ NO

AUTOMOBILE RECOVERY

ROUGH SKETCH (NOTE ANY DAMAGE TO VEHICLE)

COURT TESTIMONY **11**

PRESENTING YOUR EVIDENCE (Testimony)

There are very few guarantees in the field of investigative public safety diving. One guarantee that exists, however, is the fact that sooner or later you will be called upon to give evidence in court. As a professional, you must be prepared to present your evidence in a clear, concise, confident, and accurate manner. Anything less is simply not acceptable.

The majority of public safety dive teams are not trained law enforcement personnel; hence, their experience with the courts is limited or totally lacking. For this reason, this chapter will be dedicated to preparing the public safety diver for the inevitable court appearance.

Giving evidence as a professional is not an easy matter. Your evidence will be subject to a great deal of scrutiny and may be challenged every step of the way. For this reason, preparation is not only important — it is mandatory. The details afforded in this chapter will enable the public safety diver to understand the court proceedings and begin preparation, but it should be remembered that a great deal of preparation must come from experience. The public safety diver who has been summoned (or subpoenaed) to court should attempt to attend court proceedings prior to giving evidence for the first time. This will at least disperse some of the mystery of court proceedings and afford a familiarity that, if nothing else, will give the diver a degree of confidence.

DRESS AND DEPORTMENT

"You never get a second chance to make a first impression."

This simple statement never applied to a situation more directly and accurately than to a court appearance. As you enter the courtroom, the judge's,

jury's, and attorneys' first impressions have a definite effect on how you will be treated and how your evidence will be accepted. Whether this is fair or not does not matter; it is true.

Both men and women should wear attire that is appropriate and professional. Proper attire enhances credibility and reflects professionalism.

Often public safety dive teams are tempted to wear team attire to court. While this may reflect a great deal of pride in their dive team, members' credibility and professionalism will be judged by their evidence — not the patches on their jackets. Avoid wearing casual attire to court.

Occasionally the witness may be called to court directly from his occupational workplace and/or simply does not possess dress clothes. In such cases, the prosecuting attorney should be advised and an apology or explanation can be made to the court prior to calling the witness. This simple act will at least show the court that the lack of appropriate dress was not a matter of neglect or disrespect.

Evidence is not always black and white. Excellent dress and deportment of the expert witness may often present the court with a credible image. This image alone will often sway the court's opinion toward acceptance. Not only the evidence is being assessed; the individual presenting the evidence is also being judged.

If the first impression of a witness is one of professionalism, the signal is clearly given that this individual and his evidence should carry weight.

PRETRIAL INTERVIEW

An expert witness should never consider giving evidence without first attending a pretrial interview with the prosecuting attorney. The purpose of the pretrial interview is to review the evidence to be presented by the witness, discuss any weak areas that may surface, and prepare both the witness and the prosecuting attorney for potential discrepancies with other witnesses and/or evidence. In most cases, a trial will involve conflicting evidence and testimony. This is simply as a result of the prosecuting attorney's attempting to present a specific scenario or sequence of events, and the defense attorney offering a different possibility.

Pretrial interviews are often conducted with witnesses in a group setting. If you have any concerns that you feel should be kept confidential, privacy should be requested. As a witness in a pretrial interview, your evidence is of importance to you and to the prosecuting attorney — not other witnesses.

At the termination of the pretrial interview, the witness should have a good understanding of what questions will be asked by the prosecuting attorney, and the attorney should know what to expect from the witness. There should be no surprises during the trial. A pretrial interview should never be considered an option. It is a standard procedure for the detailed preparation of any court case and serves only to prepare the witnesses and the prosecuting (or defense) attorney for an efficient presentation of evidence.

It should be noted that pretrial interviews are also conducted with lawyers other than prosecuting attorneys. In civil proceedings, coroner's courts, and civil proceedings, both sides conduct pretrial interviews. On occasion, even in a criminal matter, the public safety diver may be called upon to give evidence for the defense. This is usually a rare occurrence, since most often the diver is called upon to present evidence he has found. When no evidence is located, however, the defense attorney may (at his discretion) subpoena the diver merely to accentuate a lack of evidence against the accused.

COURT CUSTOMS

An integral part of the pretrial interview, especially for the public safety diver who has little experience at giving testimony, is understanding court proceedings and customs. Since the uninitiated knows little of what to expect, appropriate questions are usually not asked and small mistakes are made that may ultimately prove embarrassing to the witness. These mistakes may be avoided by attending at least one trial in the court where evidence will be given. This will familiarize the witness with the physical layout of the courtroom as well as other details such as the swearing-in procedure.

There are many customs observed in various courts and countries. In many places it is appropriate to bow (slightly) toward the judge as a display of respect. This bow is performed upon entering or leaving the courtroom. Another custom that is often observed is the kissing of the Bible during the swearing-in of the witness. When this is required, the Bible is merely placed against the lips (open or closed) at the end of the oath. Other customs that should be known include whether the witness should sit during testimony or stand. In some courts, a seat will be offered, but standing conveys a degree of formality and respect for the court, which is appreciated. At least the voice will carry better when standing. The judge and jury always appreciate a clear, well-projected voice that ensures your evidence is understood clearly the first time.

Virtually all judges take notes while verbal testimony is being given. Always be aware of this, and pace your testimony with the judge. When giving evidence and it is noticed that the judge is writing, it is not only permissible, but polite, to pause until the judge looks at you. This means he has caught up and you may proceed. If there is doubt, the professional witness will simply ask: "Your Honor, I believe you are taking notes, am I talking too fast?" Remember, you are giving evidence to the court. It is your responsibility as an expert witness to give your evidence in such a manner that the court can understand and record it.

Court reporters usually record verbal testimony in shorthand. This is done either manually (pen and paper) or via a typewriter-like machine. In most cases, their records are backed up by magnetic tape recordings. If the court reporter is not able to keep pace with the testimony, it is his responsibility to signal the judge, who will in turn ask the witness to talk more slowly or speak louder. Speech should be slow, clear, and loud enough for all to hear. Verbal testimony that is low in volume, halting, or unclear not only becomes difficult for the court to interpret, but also conveys an air of incompetence or a witness lacking confidence. Neither is acceptable in an expert witness.

An additional custom that should be known prior to entering court is ascertaining the "title" of the Judge. Depending on the country and the court, titles range from "Your Honor" to "My Lord" (M'Lord), "My Lady" (M'Lady), etc. When in doubt, the least respectful term, "Sir" or "Madam," will usually suffice. Knowing how to address the court properly, however, is the sign of a seasoned professional and will quickly assist in gaining credibility.

EXPERT WITNESS

The ultimate goal of the public safety diver is to be accepted by the court as an expert witness. While this may sound incredibly difficult, it is not.

Depending on the country, the jurisdiction of the court or even the court's (judge's) own preferences, the conditions necessary to be declared an expert witness may vary.

An expert witness is a person who has proven skills, education, experience, abilities, and/or knowledge in his field of expertise or testimony that is not possessed by the average person. This is the common definition of an expert witness. The question then arises as to how much more than an average person must the expert witness prove he possesses? The answer to this question lies solely with the court and will usually vary.

Simply, the higher the court and the more serious the case is, the greater the amount of knowledge the witness must possess. A great deal of credibility will also be established by having been declared an expert witness in a lower court.

The public safety diver who wishes to be declared an expert witness must prepare carefully. This preparation must begin well in advance of any potential court appearance. The importance of being accepted by the court as an expert witness cannot be overstated, since it invariably affords credibility to your testimony.

In most cases, it is a simpler or easier matter to be declared an expert witness in a lower court; the best approach would be to seek permission from local authorities to be allowed to give evidence in one of the lower courts. Coroner's courts afford this opportunity and are to be considered a lower

court since the coroner's court does not possess the authority to ascertain guilt or punish an individual. The coroner's court is also a court that would readily seek expert evidence to explain even a simple drowning. Rules of evidence and testimony procedures in coroner's courts, while they must conform to a specific standard, are on the whole more relaxed and forgiving when compared to those of a homicide trial held in a higher court.

The coroner's court then becomes an excellent and relatively safe venue for the public safety diver to begin his experience as an expert witness.

It is only fair to clarify that while coroner's courts are considered a lower court, and their rules of testimony may be somewhat relaxed compared with a higher court, the coroner's court fulfills a very important social and legal need. The public safety diver who seeks permission to give evidence in a coroner's court should never take this duty lightly. For the public safety diver, the coroner's court provides not only a proving ground, but fulfills an important niche in our social and judicial system.

The mechanics of applying to the court for acceptance as an expert witness are simple. After being sworn in, the witness is asked (by previous arrangement with the prosecuting attorney) to relate his education, activities, and experience in his area of expertise. The witness will then begin a synopsis of his life's experiences and education, and at the termination of this, the prosecuting attorney will ask for the court's permission to have the witness accepted as an expert in his field. The opposing counsel will have the opportunity to agree or object, and the court (judge) will render a decision.

In seeking acceptance as an expert witness, caution must be exercised in the selection of a title for the field of expertise. The safest field of expertise would be to consider a title such as public safety diving. This allows a broad area of evidence to be admitted safely. Generally, it is not asked that a public safety diver be accepted as an expert witness in such subjects as postmortem physiology (unless he were a pathologist), death/homicide investigation (unless he were an experienced police investigator), scuba fatality investigation (unless he were an instructor and investigator), or even aquatic air-crash investigation (unless he had training in that area of expertise). If acceptance is granted, and the witness is declared by the court to be accepted as an expert witness in the field of public safety diving and investigative underwater procedures, he is then able to offer testimony on a wide range of diving- and recovery-related topics.

Caution: During the court proceedings, even after being accepted as an expert witness, an attempt may be made to discredit the witness by having him either agree to, disagree to, or offer testimony in an area he is unsure of. The hallmark of an expert witness is the ability to honestly declare to the court, "That falls outside of my area of expertise." A frank admission to the court will always establish credibility to the witness. A successful expert witness may have his evidence questioned, doubted, or even accused of being incorrect but never proven wrong or altered.

When giving evidence, perhaps the most important rule to remember is, "A witness, once impeached, becomes useless." Simply put, if a witness overextends his ability on the stand, he may fall off and not be allowed back on.

Importance of Being Accepted as an Expert Witness

The greatest single difference between a witness and a witness who has been accepted as an expert is that expert testimony not only conveys a greater degree of credibility, but it also affords the witness the option of offering opinion. Opinion is always subject to scrutiny by the courts, even when offered by an expert, but in many cases expert opinion has tipped the scales of justice in the direction of a conviction or acquittal.

In offering an opinion, the expert witness must be careful to clarify to the court that it is indeed only his opinion. A brief test of expert opinion would be to ask the question, "Would the vast majority of other professionals in this field agree with me?" The answer of course, should be yes. If the answer is no, the opinion must be substantiated with reasons based on specific experience, knowledge or education that is relevant. It must be remembered that expert opinion is usually offered to nonexperts. Testimony should be given using simple language.

The goal is simply to convince the court that your opinion is the correct opinion; hence, your testimony should be simple, logical, and convincing.

BEING SWORN IN

In most courts, the witness will be given a choice. Since not all witnesses are of the Christian religion, they may elect to be sworn in according to their own religious customs. If such be the case, the district attorney should be notified prior to the trial so that arrangements may be made.

In situations where the witness professes to be agnostic or atheist, he may simply take an affirmation. Affirmations are usually closely worded to the standard oath, omitting any reference to God. While it is not openly admitted, it is generally held that evidence given under oath is more credible than that given under an affirmation. The taking of an oath is a serious matter. It implies a promise of truth and acknowledges both temporal and spiritual punishment if the oath is not adhered to.

The professional witness will tell the truth in all cases — even when his evidence does not assist in the objective, which may be obtaining either a conviction or an acquittal.

Not only is it morally, legally, and spiritually wrong to lie under oath, any attempt to do so (or even the appearance of doing so) will seriously jeopardize any further testimony. Remember, "A witness, once impeached, will be deemed unreliable."

THE TRIAL

The trial and your appearance therein as a witness is the culmination of all your training, experience, and work.

The court transcript presented in this chapter is characteristic of the type of evidence/testimony the investigative public safety diver may be called upon to give. The details in this instance are from an actual homicide. The transcript has been shortened and paraphrased only for the sake of brevity and clarity. Notes in brackets and italics explain what is happening behind the scenes.

On the Scene: A Case of Murder (or an Accident?)

This incident (homicide) involved the recovery of the body of a young woman from a river. The body was located approximately one-quarter mile (400 m) downstream from a small point of land that protruded into the river. This point of land was referred to as "Brewster's Bar," chiefly because it was a favorite drinking and party location for local youths.

The night prior to the body recovery, Brewster's Bar was the site of a high school graduation party. Approximately 50 youths had attended the party, and liquor was consumed freely there. Because this site was easily accessible to the public, and memories were somewhat distorted due to the consumption of alcohol, an accurate attendance list was never constructed.

No one had noticed when or how the victim, Lucille Bertrand, had left the party. She was last seen at approximately 1 a.m. Her body was discovered later the same morning by a local resident, who was walking his dog along the river bank.

It would have been easy for the police to assume this was a simple accidental drowning; however, they did not. The public safety dive team was called in to recover the body, even though it could easily have been removed with a 5-foot (1.5-m) pike pole. The dive team was also asked to search for any evidence that would help to explain the cause of what was thought to be a routine drowning.

Prosecution

It was the opinion of the prosecution that the individual who was charged with the homicide left the party at Brewster's Bar with the victim. After attempting to force himself on her and she resisting, a fight ensued. Lucille Bertrand was beaten and choked to death, then sexually assaulted. Her body was thrown into the river, and the accused sped away in his car.

Defense

It was the opinion of the defense that his client "may or may not" have left the party with Lucille Bertrand, but that she later returned to the party, went swimming and drowned. Her body was washed downstream to where it was finally located.

Transcript from a Murder Trial

The following abbreviations appear throughout:

D.A. — district attorney (prosecuting attorney)

C.A.C. — court-appointed counsel (the accused's lawyer or attorney for the defense).

Metric conversions for distances/depths included in this transcript were not stated during the trial. They have been added herein for informational purposes only.

The trial has begun, and all preliminary functions have been completed. The jury has been selected and sworn in, the charge read, and the district attorney has sought permission from the judge to call the first witness.

TRIAL TRANSCRIPTS

Speaker	Statement(s)
Judge:	Mr. Howe (court-appointment counsel for the defense, or C.A.C.), are you and your client ready to proceed with this matter today?
C.A.C:	Yes, we are, Your Honor.
Judge:	(Nods to the D.A.) Then you may proceed. Call your first witness.
D.A.:	Call Corporal Robert Teather to the stand, please.
Judge:	Do you wish to be sworn or affirmed?
Witness:	Sworn, Your Honor.
Judge:	(Nods to court bailiff.) Proceed, please.
Bailiff:	Do you sear that the evidence you shall give touching the matters in question will be the truth, the whole truth, so help me God? [*The actual oath may vary in different court jurisdictions.*]
Witness:	I swear.
Judge:	You may kiss the Bible… you may sit if you like. [*The custom of kissing the Bible also varies.*]
Witness:	Thank you, Your Honor, I'll stand.
Judge:	As you wish.
D.A.:	Cpl. Teather, would you please identify yourself to the court?
Witness:	My name is Cpl. Robert Gordon Teather, T-E-A-T-H-E-R. I am a member of the Royal Canadian Mounted Police, presently stationed in the city of Vancouver, Province of British Columbia. I have been a member of the RCMP for approximately 20 years.
D.A.:	And were you so employed on the 20th of June 1985?
Witness:	Yes, I was.
D.A.:	Your Honor, if it please the court, prior to Cpl. Teather beginning his testimony or entering any evidence, I would ask that his résumé be reviewed and his credentials assessed. I am asking, Your Honor, for the court's permission to have this witness declare as an expert in the fields of public safety diving, drowning accident investigation, river currents and the recovery of bodies from an aquatic environment.
Judge:	That's a big area… proceed.

D.A.: Cpl. Teather, would you please explain to the court your training, experience, and education in this activity and areas that have just been mentioned?

Witness: I have been involved in sport diving for approximately 25 years...since I was 13 years old. In the past 20 years my activities have included being certified as a sport diving instructor with an internationally recognized agency as well as conducting several Underwater Recovery courses for the RCMP and other police and fire departments in Canada and the United States of America. [*The witness, when seeking to be accepted as an expert, should come well prepared. A résumé and full details regarding training, education, and expertise should be available to the court.*]

I have listed these in my résumé. I have also instructed specific topics of public safety diving, as is listed in the résumé in...uh...lectured and taught in the specialties of river recoveries, ice diving, body recovery, postmortem physiology, scuba fatality investigation, swimming fatality investigation, aircraft and automobile recovery, firearms recovery and preservation, and the handling and preservation of other physical evidence. In addition to this, I have instructed in the use of various types of land and underwater cameras, and underwater and land crime scene photography. I am accepted as a police diving instructor by the RCMP and am presently in charge of a police dive team that serves the Province of British CDA. In addition to these activities, I have a list...here it is.. on these pages is a list...a bibliography of all the courses I have taken and all the books I have studied. All these books are in my possession at home in my library. To date I have been involved in more than 200 police diving operations, each one ranging in duration from under one hour to three months. I have recovered approximately 100 bodies. In addition to these activities, I.... [*This "list" should not be padded with meaningless information, but should be given slowly and clearly. Eventually, counsel for the defense may tire and agree. Clearly, experience is the greatest qualifier in a field such as this. Education must be from an accredited agency.*]

C.A.C: Your Honor, I have no objection to this witness being accepted as an expert in his field... if it please the court. [*Counsel for the defense interrupts and agrees. He does not want the list of qualifications to be too long.*]

D.A: Your Honor, would you like Cpl. Teather to continue?

Judge: I see by his résumé he has been accepted as an expert in his field many times...I certainly have no problems with this witness being called an expert witness.

D.A: Thank you, Your Honor. Cpl. Teather, would you please relate, to the best of your recollection, your involvement in this matter that is presently before the court?

Witness: Your Honor, may I refer to my notes? [*Permission must be sought prior to referring to notes.*]

Judge: When were your notes made, Corporal? [*All notes must be made at the time of the occurrence or shortly thereafter.*]

Witness: My notes began with a telephone call I received at 7:35 a.m. on the 20th of June 1985 and continued until I left the dive site at 7:45 p.m. My notes, Your Honor, were made...uh...as I made my observations...well, except for when I was diving. I made notes on those observations immediately afterward.

Judge:	Immediately?
Witness:	Within about 10 to 15 minutes, Your Honor.
Judge:	Well, I certainly have no problems with you referring to your notes, but I would ask that you exhaust your memory first. This is common procedure.
Witness:	Thank you, Your Honor.
D.A:	Would you continue, please?
Witness:	On the 20th of June 1985, I arrived with two other members of our dive team at a location on the Fraser River called Brewster's Bar. Brewster's Bar is a sand bar...a peninsula of land on the south side of the Fraser River, situated approximately one mile [1600 m] west of Jefferson Road. This location is in the Municipality of Langley, and the Province of British Columbia, Canada.
	Am I going too fast, Your Honor? [*With little or no prompting, the witness must now tell a story of his involvement. Date, time and location must be given. The location determines whether or not this court has jurisdiction.*]
Judge:	(Holding up his index finger) If you would just slow down a bit, please.
Witness:	Sorry...Well, we arrived at Brewster's Bar at approximately 8 a.m. and were led to a site approximately one-quarter mile [400 m] downstream. The investigator, Sergeant Calland, had told us that....
C.A.C:	Objection, Your Honor. This witness is about to offer hearsay evidence. The witness can relate only what he saw and did. [*Second-hand information is hearsay and is not admissible.*]
Witness:	Sorry, Your Honor. [Apologies to the court are in order.]
Judge:	Continue.
Witness:	I attended the site...one-quarter mile [400 m] down-river from Brewster's Bar, and from the embankment I could see what appeared to be a body, a human body in approximately 4 feet [120 cm] of water, just about 3 or 3 feet [90-120 cm] from shore. The water was murky, and I could just barely make it out. We...uh...I took about 14 minutes to inspect the shoreline, took several photographs, then I suited up and entered the water.
D.A:	Please go on.
Witness:	Like I said, I was told...I...uh...I entered the water approximately 40 feet [12 m] downstream from what appeared to be a body, and slowly worked...swam underwater up to it. What I found was the body of a young lady lying prone...face down...in about 4 feet [120 cm] of water. She was dressed...or partly dressed in a bra, blue jeans and a blue high-topped running shoe on her right foot. Prior to moving the body, I photographed it using an underwater camera. After taking the first series of photographs, I continued to inspect the body. [*The witness finds himself entering into hearsay, stops momentarily, then resumes.*]

D.A:	Please tell the court exactly what you saw.
Witness:	The body appeared to be that of a 17- to 20-year-old female. A very attractive lady. *[The prosecuting side of this case is attempting to paint a picture of the victim as the "pretty girl next door." This is a routine procedure to elicit emotion from the jury. Counsel for the defense will object, but if done properly and with good taste, it is admissible to describe the victim's appearance.]*
C.A.C:	Objection, Your Honor.
Judge:	On what grounds?
C.A.C:	The witness is giving a subjective opinion in an attempt to sway the jury.
D.A:	Your Honor, if I may?
Judge:	Go ahead.
D.A:	Cpl. Teather is not qualified as an expert witness with regards to human appearances, but I believe he is indeed human. If it please the court, we will be introducing photographs later, and I submit that Cpl. Teather's observation, and I quote, "a very attractive lady," is an accurate representation, and while this may or may not be critical evidence, it is an accurate one and I submit he is qualified to judge attractiveness, Your Honor. Cpl. Teather is merely trying to accurately describe what he saw. Would we exclude his evidence on a tree if he described it as a big, healthy tree?
Judge:	Indeed. Objection denied. This witness is using a term he believes to be accurate and descriptive. Go ahead, Corporal, but I might caution you to keep your descriptions to a minimum.
Witness:	Yes, Your Honor. Uh...where was I...
Judge:	In 4 feet *[120 cm]* of water, looking at an attractive...a very attractive lady. *[It is not uncommon for the judge to help put a witness at ease.]*
Witness:	Yes...well...after taking several photographs, I slowly rolled her over. I noticed she was wearing an earring...her ears were pierced...she had what appeared to be a laceration...a cut on her lower lip...to the left of center. She was wearing a watch, a Bulova, on her left wrist. It was flooded with water and was stopped at 5:30 o'clock. It was an analog watch, Your Honor... no indication of a.m. or p.m. *[Observations are given slowly and in the order they were made.]*
Judge:	Thank you. Did it have a second hand?
Witness:	Uh...let me think...I don't recall, Your Honor. May I refer to my notes? Witness seeks permission to refer to notes.
Judge:	Yes, please do.
Witness:	Yes, Your Honor, the second hand was stopped at the "2" position on the face of the watch, and the watch recorded the time as 5:28.

Judge:	You may keep your notes out if you like.
Witness:	Thank you. I checked...I, uh, continued to make my observations of the body of this young lady and noticed several scratches on the side of her throat and neck. The left side...that is her right side...she had three small scratches on the side of her throat...uh...her neck, just below her right ear and one gouge just on the other side of her...what would be her windpipe... her larynx. Her eyes were half-open; the corneas of both eyes were clear. I removed my rubber diving glove and parted her lips approximately one inch [25 mm]. There were no signs of any bottom sediment or material inside her mouth. Her jaw was partly open and I...well, I closed it. I could feel only a slight resistance in manipulating the jaw.
	I then placed a small plastic bag over the head of the body, carefully squeezing out most of the water, and I secured it around the upper portion of her neck with a rubber strip equipped with a Velcro closure. Not very tight, just enough to keep any materials in or on the head from being lost.
	I bagged both hands in a similar manner, but that was a few minutes after she was removed from the water. Sorry, I'm jumping ahead of myself. As I continued to inspect the body, I noticed she was wearing a white bra, with no other clothing above the waist. The bra had been fastened, but both breasts were...uh...well, they weren't...covered. The bra had been pulled.... [Observations are detailed, objective, and clear.]
	[Rigor mortis was not fully established in the small jaw muscles. The victim's mouth was later found to contain semen.]
C.A.C:	Objection, Your Honor. Objection. How does the witness know the young lady's bra had been pulled up? Is there any proof or indication that she hadn't just put it part-way on? [Counsel for the defense recognizes that a "violent" scenario is being described and is attempting to stop the testimony. In doing so, he merely draws attention to it.]
Judge:	Sustained. Corporal, do you have any great expertise that qualifies you to say how a woman puts on her bra?
Witness:	No, Your Honor. I'm sorry, I meant to say merely that her bra was fastened in the back, with both cups worn or raised...in place above her breasts.
Judge:	Thank you. Please continue.
Witness:	Well, the body was removed from the water at 9:05 a.m., and in company with Sgt. Calland and another man who was never identified to me, they left in the coroner's vehicle. We...I remained at the site.
D.A:	Could you please tell the court if you conducted a further search of the scene, and if so, what you found?
Witness:	Our first duty...sorry, Your Honor...My first action was to complete a search of the immediate site, both on land and underwater with a metal detector. Your Honor, may I refer to the photographs I brought with me today?
	[Photographs are primarily utilized to make verbal testimony easier and clearer for the court to understand. Photographs place everything in perspective.]

Judge:	Who took these photographs? [*Photographs are introduced by the photographer.*]
Witness:	I did, Your Honor.
Judge:	Do they represent a fair and accurate representation of what you saw on 20 June 1985? [*This question must always be answerable in the affirmative.*]
Witness:	Yes, Your Honor, they do.
D.A:	Your Honor, this witness has brought eight booklets with him. I would ask that one copy be marked exhibit #1 and the other copies be distributed to the members of the jury and, of course, one copy for my learned friend, counsel for the defense. [*Photographs should be set in booklets, labeled and in sufficient quantity for the jury.*]
Judge:	Thank you, Your Honor. I'm sure my learned friend has no objection. [*In many jurisdictions, lawyers refer to each other as "my learned friend."*]
C.A.C:	No, Your Honor. I've already seen these photographs. I have no problem with their admissibility.
Judge:	Exhibit #1. Go ahead, Corporal.
Witness:	In photograph #1, you will see a shot...uh a view of the scene. Looking east, that's up-river. In the distance you can see a group of trees jutting out from shore into the water. That is called Brewster's Bar...it's almost an island except for a narrow land-bridge about 20 feet [6 m] wide that connects this raised sand bar to the mainland. This semi-island, Brewster's Bar, measures about 300 yards [300 m] long...that is, east to west, and about 100 yards [100 m] in width.
	Photograph #2 is a closer view of the area where the body was recovered. The red buoy marks where the body was found. If you look carefully, you will be able to make out the outline of her body, Your Honor. It's sort of like a light shape directly below the buoy. [*The photographs are explained in the order they appear in the booklets.*]
Judge:	Yes, I see it. (Turns to jury.) Is anyone not able to make that out? Continue, please.
Witness:	Photographs numbering 3 through 7 were taken underwater and are close-ups of this lady's body before it was moved. In particular, photograph #3 is a full-length photo from the... of her right side. Almost full-length. #4 from the perspective of her head looking down the body toward her feet, #5 from her left side, and #6 was taken from her feet looking up toward her head, showing her shoe. #7 was taken looking directly down the body... it shows the back of the head, neck and shoulders of the victim. This photograph was taken parallel to the body axis, and visibility was so poor that her entire body could not be clearly seen.
	Referring now to photograph #8, this is a photograph of the scene, taken from shore... from the same location as #2 was taken. In the lower left corner, a wooden stake with a small piece of orange surveyor's flagging tape is visible.
Judge:	Yes. I see it.

Witness:	That is the location where we found an earring, 8 feet [2.4 m] from the water's edge.
Judge:	You found an earring in the sand?
Witness:	Yes sir.
Judge:	Continue.
Witness:	Out in the water, approximately 18 feet [5.5 m] past the first orange buoy marking the body, is a second buoy...a white marker. This white buoy marks the location of underwater photograph #8. Photograph #8 was taken in 6 feet [2 m] of water and shows a blue high-topped running shoe.
D.A:	Excuse me, Corporal, but I would like to interrupt. The earring you located...in the sand, could you describe it?
Witness:	Yes. It is a stud-type earring, designed for wearing with pierced ears. It is gold-colored and has a pearl or pearl-like object in the center with what appears to be diamond chips surrounding the center. [*Without proper qualifications, the witness cannot identify the stones as diamonds, etc.*]
D.A:	And could you produce that earring today? [*The earring is admitted as evidence.*]
Witness:	Yes, I brought it with me. Here we are.
D.A:	And has this earring been in your possession since you found it?
Witness:	No, sir.
D.A:	Then could you please tell the court how it is that you know this to be the earring you seized or found at the scene, on the 20th June 1985? [*Unless an exhibit has been in the sole custody of the witness, it must be identified as the object mentioned by some specific marking or characteristic. When it cannot be marked, it should be packaged and labeled.*]
Witness:	The earring was picked up from the sand by myself...after locating it with the metal detector. I immediately looked at it...I, uh...I went to the back of our truck, removed a small exhibit bag, and I placed the earring inside the bag.
	Prior to this, the bag was clean and contained no foreign objects or material that I could see. After placing the earring in the bag, I removed the backing from the flap portion of the bag and sealed the bag by folding the prepared sticky flap tightly over the neck of the bag. On the flag in my handwriting is "R.G.T. 2611220 June 85." Referring to my notes, your Honor, I have it recorded that this earring was located at 11:47 a.m., sealed in the exhibit bag at 11:53 a.m., and later turned over to Sgt. Calland at 5 p.m.
	[*Initials, police officer's number, and date. The history of the earring is given along with the name of the person who next took it into possession. This is the first step in proving continuity of an exhibit. Sgt. Calland must later corroborate this testimony, agreeing to receiving it and following through with what he eventually did with it.*]

D.A.: And where was that done?

Witness: At the site. Sgt. Calland had returned to the site later in the day.

D.A.: Exhibit #2, if I may, Your Honor?

Judge: Granted...Exhibit #2. May I see that after it has been entered?

D.A.: Cpl. Teather, would you please tell the court if you seized any other items from the scene?

Witness: Yes. The blue high-topped running shoe pictured in photograph #8.

D.A.: And what, if anything, was familiar about this shoe?

Witness: This running shoe was located approximately 15 feet [4.6 m] out past the body was the same... uh.... make... brand name, style, color and size as the one worn by the body of the young lady...with one difference.

D.A.: And that difference was?

Witness: She was wearing a right shoe, and this was a left. [Suggesting this running shoe belonged to the victim.]

D.A.: And do you have this shoe with you today?

Witness: Yes. I have it here.

D.A.: Would you produce it, please?

[This exhibit was marked by the witness. It was later turned over to Sgt. Calland at the same time as the earring.]

Witness: This is the shoe I...uh...this is the running shoe I recovered about 15 feet [4.6 m] away... north of the body. In permanent marker I have written on the side, "R.G.T. 26112 20 June 85" and, referring to my notes, Your Honor, this shoe was removed from the water at 12:16 p.m. on 20 June 1985.

D.A.: And did you do anything to this shoe when it was in your custody?

Witness: Yes. Using a needle and thread, I passed a piece of black thread directly through the knot...the bow of the shoelace, so it could not be untied. The shoelace was later found to be tied with a left-hand bow, while the shoe she wore was tied right-handed. [The inference is that someone else tied this shoelace.]

D.A.: Then the shoelace was tied when you found it?

Witness: Yes, sir.

D.A.:	I notice the lace is no longer tied.
Witness:	But you did not untie it?
Witness:	Correct.
D.A.:	Do you know how or why it came to be untied?
C.A.C:	Objection, Your Honor. The district attorney is attempting to lead this witness into hearsay. [*The witness cannot give evidence as to how the shoelace was examined unless he was present. This would be hearsay evidence.*]
Judge:	Corporal, were you present when the knot was untied?
Witness:	No, Your Honor.
Judge:	Sustained. Please continue.
D.A.:	Would it be fair to say that this shoelace was later examined at the crime laboratory?
C.A.C:	Objection again, Your Honor. Unless this witness was present, or at least transported the shoe to the crime laboratory, he cannot give evidence as to its subsequent examination, and clearly he turned this shoe over to Sgt. Calland at 5 p.m.
Judge:	Sustained. Would the district attorney please continue along a different line of questioning. I have made a footnote, if you'll pardon the expression, that this is the second time you have attempted to solicit hearsay evidence from this witness.
D.A.:	Sorry, Your Honor. I'd like to enter this shoe as Exhibit #3, if it please the court. Cpl. Teather, I notice in photograph #2 what appears to be a long mark on the gravel shoulder of the road. Would you comment on this, please?
Witness:	The mark measured almost 8 feet [2.4 m] long and 6 to 12 inches [15-30 cm] wide, and seems to have been made by a wheel of a car as it accelerated on the pea-gravel shoulder of the road. [*Inference is being made that a vehicle sped away from the scene.*]
D.A.:	And why do you say that?
Witness:	The shoulder of the paved road appears to have been freshly graveled...the mark was like one would make when accelerating away...the fresh gravel was removed and that is in line with the black mark on the pavement.
C.A.C:	Objection, Your Honor. This witness is giving evidence out of his area of expertise... how can he tell it...this shoulder was freshly graveled? [*Evidence such as this should be reviewed with the D.A. in the pretrial interview so that he can prepare to overcome any such objection.*]
D.A.:	Your Honor, I think it fair to accept that a police officer with 20 years' experience can judge whether or not the shoulder of a road was freshly graveled or not.
Judge:	I think so, too. Objection overruled. Please continue.

Witness: As I was saying, Your Honor, the shoulder of this portion of highway...roadway appeared to be freshly graveled, and what led me to believe the long mark where the gravel was removed was caused by the wheel of a car was the corresponding tire skid and yaw mark, which is quite visible in photographs #1 and 2. Also, there was similar fresh gravel strewn in the river... on the bottom...for a distance of 10 to 12 feet [3-3.6 m] directly in line with this mark.

[The witness was prepared to say that the river bottom is composed of fine sand, but was never asked.]

D.A: Cpl. Teather, your expertise in this field of public safety diving, drowning accident investigation and body recovery has been accepted by other courts as well as this one, so I would like to ask you for your opinion. [Opinion may be sought from an expert witness.]

Witness: Yes?

D.A: Did Lucille Bertrand...the lady whose body you recovered...did Lucille Bertrand drown... In your opinion...and please give your reasons for coming to any conclusions you might offer this court.

Witness: In my opinion, Lucille Bertrand did not drown, or at least if she did ultimately drown, this was not an accident. There were just too many discrepancies, and I would like to list them.

First, the temperature of the water was 59°F [15°C], not a comfortable temperature for most people to go swimming. Second, the body... her body was found in water shallow enough for her to easily stand in.

There were other observations that I made... observations that are not usually seen in accidental swimming drowning scenes. The lady's dress was not normal. She had one shoe missing and her bra was pulled... sorry your Honor...her bra was... uh...worn up over her breasts. She was missing one earring, showed signs of physical trauma to her face... her lip, that is... and her neck. There were no large rocks in the area to fall on...her. A similar shoe was found out in deeper water. It could not have drifted there in that current, and the gravel I spoke of that was removed from the shoulder of the road... [When opinion is given, it must be given clearly — in simple terms and substantiated by logical observations that lead to a logical conclusion. In short, an opinion must be explained in terms the court can comprehend.]

Two pieces were found on her...one on her back and one on the back of her neck. Photograph #7 shows these two pieces of pea gravel similar to that which covered the shoulder of the road. Photograph #2 will show, if you look at the marker used to indicate the location of the body, it is lying directly in line with the mark in the gravel and the black mark on the asphalt.

In most cases, or rather many cases of accidental drowning, sand, mud or other bottom debris is often present in the mouth of the victim. I saw none in her mouth.

Putting it all together, the physical trauma, missing earring, shallow water... shallow enough to stand in, its temperature and the way the body of Lucille Bertrand was dressed, a typical swimming accident is not a conclusion I would come to.

D.A:	And your conclusion is?
Witness:	Simply that Lucille Bertrand was at very least unconscious when she entered the water, perhaps already dead, and that she had been a fight or scuffle prior to that. [Conclusion is short and to the point.]
D.A:	I have no further questions from this witness, Your Honor... I would like to reserve the right to re-examine this witness at a later time, however. [Since an opinion is sought that may be argued later in the trial, the D.A. may wish to recall the witness.]
Judge:	So noted. Does the crown-appointed counsel for the defense have any questions he would like to put to this witness?
C.A.C:	I do, Your Honor.
Judge:	Proceed.
C.A.C:	Cpl. Teather, the shoe you located approximately 15 feet [4.6 m] away from the body... can you say for sure it belonged to the deceased? The counsel for the defense is attempting to disprove the shoe belonged to the victim. While ownership cannot be proved, he is failing in his attempt merely by drawing attention to the shoe.
Witness:	The chances of...
C.A.C:	Please answer the question with a simple "yes" or "no."
Witness:	Well, it was...
C.A.C:	Your Honor??? [Witness must answer the question as directed.]
Judge:	Would the witness please do as instructed. Answer the question put to you as directed... with a "yes" or "no."
Witness:	No. I can't say for sure.
C.A.C:	And the earring you recovered in the sand...using your metal detector...did it match the one worn by the victim?
Witness:	Yes.
C.A.C:	Are you a jeweler or gemologist? How do you know it was a match?
Witness:	Sorry — it was similar.
C.A.C:	Thank you. So far we only have an earring and a shoe. There is no evidence that these belonged to the deceased.
Witness:	So far.

C.A.C:	Now the deceased presumably wore a blouse, a shirt, or sweater...something on her upper body during the evening...when she was last seen.
Witness:	I don't know...I would assume that would be a safe assumption.
C.A.C:	Where was it?
Witness:	We...I never found it. [*The blouse was never located.*]
C.A.C:	And did you find anything else, either with your metal detector or your diving in the water?
Witness:	On the sand, near the water's edge we...uh, I searched the narrow 3- to 5-foot [90-150 cm] strip of sand and found more than 100 pieces of rusty iron, nails, wire, bolts, washers, etc., 23 bottle caps, three aluminum pull tabs, a quantity of metal foil...cigarette-package foil I'd guess, and seven pennies and one five-cent piece. They had been there a long time. [*The thoroughness of the search is demonstrated.*]
C.A.C:	Cpl. Teather, I put it to you that Lucille Bertrand went swimming at Brewster's Bar, which is located upstream. She was alone and drowned...the river washed her body down to the location in this photograph. [*This story is an alternate scenario the defense would like the jury to accept. Brewster's Bar was the site of a graduation party the night previous. Defense is attempting to submit the possibility that the victim was drunk, went swimming and drowned.*]
Witness:	No.
C.A.C:	Perhaps you will think not, but I suggest it is at least possible?
Witness:	No, that would be quite impossible.
C.A.C:	Well, if you think so, perhaps you could explain to the court how it is impossible for a body to float downstream in the river. [*An attempt is made to confuse the witness.*]
Witness:	That's not what I said. [*The witness clarifies his previous statement.*]
C.A.C:	Well, that's the impression I got. What are you saying, then?
Witness:	I said it would be impossible for a person to drown on the night previous at Brewster's Bar and have their body wash or drift to the location we found Lucille Bertrand.
C.A.C:	You say "impossible." Can you prove this?
Witness:	I believe so.
C.A.C:	Go ahead, then. I think the court would like to hear this proof. [*The defense will soon regret this move. Normally the prosecution would ask for this evidence, but in this case it is merely a courtroom ploy to have the defense embarrassed by helping to convict his own client. These tactics are usually discussed with the witness prior to the trial.*]

| Witness: | Myself and two other members of our dive team put on our diving gear...air tanks, suits, everything, and then weighted ourselves to approximate the weight of Lucille Bertrand's body...her underwater weight. |

C.A.C: Excuse me? How much did she weigh underwater?

Witness: Nine pounds, plus or minus a pound [4100 g, +/- 450 g].

C.A.C: And how do you know this? How can you weigh the body of an accidental drowning victim underwater?

Witness: The same way we weighed Lucille Bertrand's body...using a set of previously calibrated spring scales...attaching them to the body bag and suspending the body off the bottom. [*Witness is refusing to accept the description of Lucille Bertrand as an accidental drowning victim.*]

C.A.C: How do you know these spring scales are accurate...springs must get stiff or something, don't they? [*Any measuring device must be proven to be accurate. The accepted way is to calibrate it immediately prior to its use.*]

Witness: Immediately before weighing the body, we...I ascertained the accuracy of the scales using a series of one-pound [455 g] lead weights in the 1- to 15-pound [455g to 6.8kg] range. Our scales were accurate on the 20th of June 1985.

C.A.C: And how did you weight yourselves underwater? Did you hang each other in a body bag? [*Defense counsel is attempting to stir the emotions of the witness. The question is answered in a straightforward manner.*]

Witness: No. We weighted ourselves that day for neutral buoyancy...that is, Your Honor, a condition where a diver with his lungs half-filled with air will neither rise nor sink. We weighted ourselves, then confirmed it by entering the water. After this, we added to our weight belts the under-water weight of Lucille Bertrand's body...in this case, 9 pounds [4100 g].

Judge: Go on.

Witness:	We entered the river at several locations on and around Brewster's Bar. That island-like projection... we wanted to see if it was possible for a swimmer to drown at that location and be carried down the river. Well, in two days, three divers did a total of 20 descents, or trial runs, as we called them. This came to a total of 60 trial runs at 60 different locations on Brewster's Bar. None of these experiments brought us down to the location of Lucille Bertrand's body. You see...just west of Brewster's Bar, approximately 100 feet [30 m], there was a rather incredible collection of loosely coiled barbed wire...nothing could get by that. We almost didn't on a couple of occasions...that is, we couldn't get by it and almost didn't return, we became so badly tangled. Also, we found that if you swam out anywhere near the outermost end of this mass of entanglement, the current from the river swept you toward the center. On several occasions we could not even swim back to the shore. The current directed everything to the center of the river channel, which is about 500 feet [150 m] from shore. The Fraser River at this point is about 300 to 400 yards [300-400 m] wide. Brewster's Island, or Bar as it is called, is eroding naturally, and as it is currently situated, it deflects the current of the river away from the shore and back out toward the center.
	Well, between the barbed wire to catch us and the current further out, it was impossible for experienced swimmers or divers in good physical condition to make it to the site where the body was found...I suspect an unconscious person would not even do as well. [In relating just how the witness came to his conclusion, the tests that were performed are given in a clear, understandable manner so that the judge and jury can be shown exactly how it would be impossible for a drowning at Brewster's Bar to arrive at the site where Bertrand's body was found.]
C.A.C:	Cpl. Teather, is it true that on rare occasion, even adults may float after drowning?
Witness:	Yes, in isolated cases.
C.A.C:	Well, I suggest that Lucille Bertrand could have drowned in a swimming accident and remained on the surface, or better yet, swam to the site where her body was found...after all, your tests were done with divers weighted down to approximate her underwater weight, were you not...you didn't test for surface currents, did you?
Witness:	Yes, sir, we did.
C.A.C:	How?
Witness:	Over a period of approximately one hour on the 21 June, we threw in 200 floats from various locations on Brewster's Bar. We ranged the distance from shoreline to approximately 20 feet [6 m] out from shore. Well, all except for three, which never left the shoreline...every other float headed out to the center channel...none of them came even close to shore where we found Lucille Bertrand's half-clothed body.
C.A.C:	But, could she not have become caught in those loose rolls of barbed wire, had her sweater or blouse ripped off and her bra pulled up? I suggest that scenario is at least possible.
Witness:	Our rubber diving suits became quite damaged that day. One of them had to be reduced to spare parts. The old wire is very unforgiving and damaging. Lucille Bertrand's body had no scratches on it at all...that I could see...or photograph. She could not have gone through the barbed wire unscathed.

C.A.C:	But, isn't it even at least possible that Lucille Bertrand somehow went swimming at this place... what is it...Brewster's Bar, and her body drifted to where you found it...one chance in a million, even...
D.A:	Your Honor, my learned friend has sought the opinion of this expert witness. [The D.A. is merely re-enforcing the credibility of the witness and the tests that were performed.]
	He has scrutinized the current tests they performed, and now he suggests a million other tests should have been done. Perhaps, Your Honor, if so, then two million? Cpl. Teather has given his opinion, based on his expertise and specific tests performed on this portion of the river. I would ask that the court decide on the validity of this witness's testimony...not my learned friend with his other million hypothetical tests or suggestions.
	Clearly, Cpl. Teather's opinion was asked for and given...and his opinion only...if the counsel for the defense would like to be sworn in and give evidence in this matter???
Judge:	That is enough.
C.A.C:	Your Honor, I was only testing the confidence of this witness.
Judge:	Continue, then.
C.A.C:	Cpl. Teather, would you describe, in your opinion, how Lucille Bertrand sustained her injuries... her cut lip and scratch marks?
Witness:	Your Honor, that falls outside of my area of expertise. I am not qualified to give evidence in this area. [The hallmark of an expert witness is that he adheres strictly to his area of expertise.]
C.A.C:	Well, did the marks seem fresh?
Witness:	I can only give observations in this matter...I am not qualified, trained or experienced enough to offer an opinion. The lip was cut, and there appeared to be no swelling or bruising that I could see. [Only observations are given; the court may draw any inference from those observations they wish.]
C.A.C:	Could you call that a fresh wound?
Witness:	I'm sorry...perhaps a qualified forensic pathologist could...
C.A.C:	Surely you can at least guess as to whether or not Lucille Bertrand's lip was cut recently.
Witness:	Sorry. I have no qualifications with regards to...
C.A.C:	Corporal Teather, have you...in the line of duty, of course...ever been involved in a fight...have you ever had a fat lip?
Witness:	Yes.

C.A.C:	Then you have some idea as to how long it takes for your lips to swell up...don't you. After you are hit in the mouth with a fist? Surely you must remember?
Witness:	There is a difference. [The witness is now subtly controlling the defense lawyer.]
C.A.C:	Well, would you please tell the court just what the difference is?
Witness:	I lived — Lucille Bertrand did not.
C.A.C:	I have no further questions of this witness, Your Honor.
Judge:	Then you may step down. Since the district attorney has reserved the right to recall you, would you please remain in this building? [The judge gives final directions to the witness.]
Witness:	Yes, Your Honor.
Judge:	You are not to talk to any other witness regarding this matter.
Witness:	Yes, Your Honor.
Judge:	You may step down.
Witness:	Thank you.

TRIAL RESULTS

Throughout the trial, the defense counsel attempted to present a scenario in which Lucille Bertrand attended the graduation party at Brewster's Bar, and after having consumed a quantity of alcohol, went swimming and drowned. Her body later washed up at the location where it was found.

The evidence presented by the dive team indicated that her body had been placed in the river, either dead or unconscious, and her running shoe, which had come off in a scuffle, had been retied by the assailant (for whatever reason) and then thrown in. Further evidence of a scuffle, in addition to the marks noted on her face and neck, was her earring, which was found at the water's edge, in the sand.

The forensic pathologist gave evidence that the cause of death was "asphyxia due to manual strangulation" and that there was evidence of semen in her mouth. A classmate of Bertrand testified that she saw Bertrand and the accused running from his car, which was parked at the site of the body recovery. She thought nothing of it at the time. While no license number was given, both the accused and his vehicle were well known to the witness.

The blood alcohol level of Bertrand, which was measured at the time of the autopsy, showed a concentration of 0.03 percent.

In this investigation and subsequent trial, the evidence gathered at the scene and the subsequent current tests performed by the dive team were major factors in obtaining a conviction of aggravated sexual assault and homicide against the accused.

THE COLLECTION AND PRESERVATION OF PHYSICAL EVIDENCE

12

Physical evidence is defined as any object, material, substance or thing that can be used to indicate a past occurrence, condition or sequence of events.

The importance of the efficient collecting and handling of physical evidence cannot be overstated. Physical evidence, once passed over, may not remain for future collection. In addition, physical evidence, if improperly handled, may be altered or destroyed, resulting in its inadmissibility in court.

The purpose of this chapter is twofold: first, to describe the crime scene and the various locations where physical evidence may be found; and second, to list common types of physical evidence and offer suggestions for its safe and efficient use, collection, and handling.

THE CRIME SCENE

Most often, the crime scene is thought of as a physical location. Due to the variance of locations, in this chapter consideration will be given to two crime scenes that are most often encountered by

the investigative dive team. They are the motor vehicle and the human body. In both cases, the crime scene will include the immediate area around the vehicle or body.

This chapter has been written as a guide for the investigative dive team. This guide for the collection and preservation of physical evidence may be utilized in two ways: by referring to the crime scene (vehicle or body) or by referring to the type of physical evidence.

The crime scene (vehicle or body) illustrations depict common types of physical evidence that may be found on or near the site. Refer to the item of evidence for details regarding the collection, preservation, and/or shipping of the evidence/exhibit. The evidence in the illustrations is located in Chapter 13, "Types of Evidence."

Note: The list of exhibits taken by a forensic pathologist is usually considerably more comprehensive than this suggested list.

CRIME VEHICLE
Suspected stolen and/or hit and run

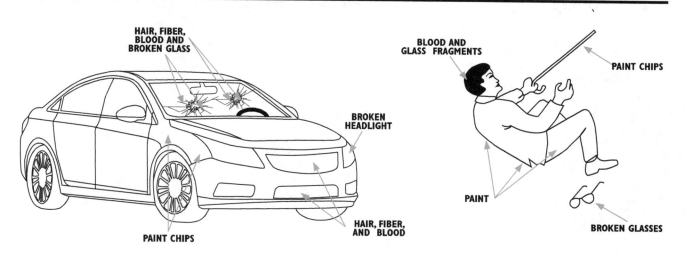

HAIR, FIBER, BLOOD AND BROKEN GLASS

BLOOD AND GLASS FRAGMENTS

PAINT CHIPS

BROKEN HEADLIGHT

PAINT CHIPS

HAIR, FIBER, AND BLOOD

PAINT

BROKEN GLASSES

CRIME VEHICLE
Suspected stolen and/or hit and run

FINGERPRINTS

TOOL MARKS

ALTERED
V.I.N. PLATE

TIRES
SLASHED OR
PUNCTURED

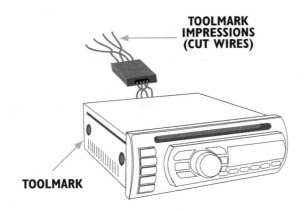

TOOLMARK
IMPRESSIONS
(CUT WIRES)

TOOLMARK

PAINT TOOLS

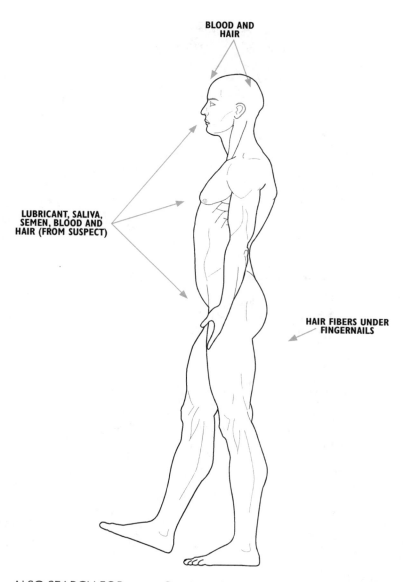

BLOOD AND HAIR

LUBRICANT, SALIVA, SEMEN, BLOOD AND HAIR (FROM SUSPECT)

HAIR FIBERS UNDER FINGERNAILS

ALSO SEARCH FOR: Condoms (semen)   Cigarette butts
                    Clothing   Paper matches
                    Buttons   Pills

"ROUTINE" AUTOPSY EXHIBITS

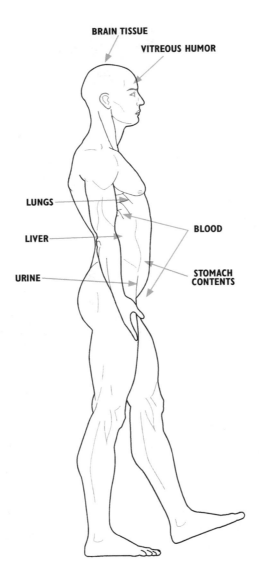

TYPES OF EVIDENCE 13

This chapter is an alphabetical listing of the most commonly found types of evidence. Each type of evidence is listed along with a commentary on its collection, preservation method, shipping instructions, and comparison standards requirements.

For a more detailed account, investigative dive teams are encouraged to consult with their local forensic pathologists and/or forensic crime laboratories.

I. AMMUNITION

Ammunition refers to live rounds, lead projectiles, and spent casings.

a. Do not mark ammunition.

b. Avoid altering or damaging ammunition during recovery.

c. Adhere to federal law restrictions when transporting or shipping live ammunition.

d. Bullets or bullet fragments may belocated in a body using X-rays. The investigator should be aware that certain types of ammunition have aluminum jackets that may not show up on X-rays if jacket core separation has occurred. Careful examination of the bullet removed from the body may be necessary to determine if all of the evidence is present. Bullets or fragments may be located in walls, soil or trees using a metal detector.

e. When retrieving a bullet or fragment from any medium, avoid the use of sharp objects such as metal forceps or tweezers.

f. Ammunition, shell casings or bullet/bullet fragments should be wrapped in a soft material and placed in a container that can be sealed and marked for identification.

2. BILE

Bile is to be collected during an autopsy if narcotics (e.g., morphine) is suspected.

a. Bile specimens should be placed in a clean, sterile container and sealed. Keep cool — do not freeze.

3. BLOOD

Blood samples/evidence is listed under the following categories:

• Alcohol Analysis

• Drug/Poison Analysis

• Blood Grouping/Comparison/DNA Typing

Alcohol Analysis:

a. If the victim is deceased, a blood sample is usually obtained during the autopsy from the femoral artery or intact heart. Obtain one or two vials using vacutainer XF947 or 10 ml. gray stoppered vacutainers contain fluoride and oxalate which are an enzyme inhibitor and an anticoagulant respectively. Mix gently.

b. Label vacutainer.

c. Refrigerate — do not freeze.

d. Transport to laboratories as soon as possible.

Drug/Poison Analysis:

a. During the autopsy, collect eight red-stoppered vacutainer vials of blood. Red-stoppered vacutainers contain no additives.

b. Blood should be collected from both cardiac and peripheral blood, when possible.

c. Label vacutainer.

d. Refrigerate — do not freeze.

e. Transport to laboratories as soon as possible.

Blood Grouping/Comparison/DNA Typing

a. If blood stain is on absorbent material (e.g., clothing), AIR DRY the stain completely and package each item separately.

b. If blood stain is on any other material (e.g., vehicle upholstery, wood, plastic), either remove a portion of the material and proceed to AIR DRY, or scrape dried blood into a vial using a sterile scalpel or knife, or using water, moisten one corner of a clean sterile piece of cotton gauze or a sterile swab. (not Kleenex or Q-tips), rub stain until cotton is dark red. If more stain remains, repeat for further samples. Once this has been done, AIR DRY the cotton.

c. If liquid blood stains are found (on rubber car floor mats, etc.), soak up the stain with a piece of clean sterile cotton until a dark red color is obtained, then AIR DRY the cotton.

d. Comparison standard. The sample may be compared to a standard (sample) of blood from the suspected source (victim). One 10 ml. purple or lavender-stoppered (EDTA) vacutainer of blood should be taken from the suspected source. EDTA is a strong anticoagulant.

e. Label all samples.

f. Refrigerate, do NOT freeze, known sample (comparison standard), and transport to laboratory as soon as possible.

g. Both suspected samples and comparison standard should be packaged and transported independently so as to avoid cross-contamination.

4. BRAIN TISSUE

a. Place in leak-proof, sterile container.

b. Seal and label container.

c. Refrigerate — do NOT freeze.

d. Brain tissue is to be collected, especially if solvents such as glue, PAM, cleaning fluids, or other inhalants are suspected..

5. BUTTONS

a. Buttons may be found in vehicles, near a body, or at any crime scene. Separation from clothing may indicate a struggle.

b. Package buttons carefully to avoid loss of any threads or fibers.

c. Comparison standard to any clothing with missing buttons.

6. CIGARETTE BUTTS

Cigarette butts were collected in the past only when lipstick or other stains were evident.

Presently, with the use of DNA comparison/ matching standards, any cigarette butts should be considered as potential evidence.

a. Do not handle with bare hands.

b. Air dry the cigarette butts completely.

c. Package in separate containers.

d. Comparison standard: If DNA comparison is anticipated, saliva sample may be obtained during autopsy in addition to the victim's blood. This sample should be packaged separately from the cigarette butts to avoid cross-contamination.

7. CLOTHING

Clothing may be submitted to forensic crime laboratories and checked for the presence of hair, fibers, or foreign particles. These may, in turn, be matched to another individual or scene (car, house, work site, etc.).

a. Air dry clothing, making sure not to shake off or remove any fibers or particles present.

b. Package each article of clothing separately in paper only after allowing the article to air dry completely. Packaging wet evidence in plastic or airtight containers should generally be avoided as it will promote the growth of mold and other organisms that may be detrimental to the evidence. Avoid cross-contamination.

c. With the clothing, include a report outlining any areas of pertinent stains or suspected particles. (e.g., hair in underwear, stains on shirt, etc.)

d. Comparison standard — Refer to headings of Glass, Paint, Lubricants, Semen, Blood, Saliva, and Urine.

e. If propellant powder residue (shooting victim) is suspected, protect the bullet-damaged area from further contamination, air dry at room temperature, and package each item of clothing separately to prevent the transfer of trace evidence.

8. EXPLOSIVE DEBRIS AND POST-BLAST INVESTIGATION

a. Upon arrival the team should establish incident command (or report to command), secure the area, and control access to the site to preserve the integrity of the investigation. The injured should be evacuated and tended to by appropriate medical personnel, witnesses identified, and transient physical evidence collected. Transient evidence is evidence that can be lost over time and can include items floating in the water, present on victims, or compromised by weather.

b. Contact the nearest explosive ordinance disposal unit (civilian or military) so the site may be searched for secondary explosive devices. This should be done PRIOR to searching the blast site for evidence. Consider the benefit detection canines and electronic explosive detectors to render the area safe.

c. Take photographs and video recordings of the area (minimize the presence of scene personnel), and begin the search for debris at the seat (origin) of the blast/explosion. Search outward from the seat in a systematic manner. Consider the benefits of a systematic grid search so collected evidence and the location where it was recovered can be documented.

d. Under the direction of a senior post-blast investigator, collect debris and record where the debris was located.

e. Place debris in airtight containers immediately, and prevent cross-contamination between collected pieces of evidence.

f. Pay special attention to metal fragments, pieces of tape, wire, wrappers (either paper or plastic), clock mechanisms, fuses, battery parts, circuit boards, etc.

g. All exhibits must be in separate containers and documented as to their location so the distance from the site of the blast can be determined.

h. Conduct a comparison analysis on any suspected source (suspect or source of manufacture of the explosive device).

9. EXPLOSIVE DEVICES AND SUBSTANCES

a. When an explosive device is suspected, immediately consult with the nearest explosive ordinance disposal (EOD) unit. The suspected device MUST be rendered safe prior to inspection and/or transportation. Any explosives in the water should be handled and dealt with by EOD technicians or by divers working under the direct supervision of a certified EOD technician.

b. In most cases, no more than 50 grams (2 ounces) of explosive or suspected explosive material should be sent for analysis. Rely on information and directions from the explosive ordinance disposal unit for collecting, handling, and shipping or transporting. Some explosive materials pose a major threat in very small amounts.

c. Comparison standard any substance found at the site of a blast.

d. Comparison standard and all explosive devices and substances must be packaged (preferably double sealed) and kept separate to avoid cross-contamination.

10. FIBERS

a. If on a movable object or garment, package the item in a sterile container.

b. If the item the fibers are found on cannot be submitted (e.g., an automobile seat), carefully remove suspected fibers and place in a clean vial or container.

c. Another (less desirable) technique is to remove fibers using sticky tape pressed against the material. The tape should be placed sticky side down on a clean plastic sheet. Label the sheet.

d. Comparison standard any suspected source of the fiber that has been collected. This includes carpets, upholstery, blankets, clothing, rope, etc. Package comparison standard carefully and separately to avoid any cross-contamination.

11. FINGERNAIL SCRAPINGS AND CLIPPINGS

a. In the past this was of limited forensic value, but with recent progress made in DNA comparison/matching techniques, it has resurfaced as a viable exhibit.

b. Fingernails should be clipped as far down as possible.

c. Scrapings may be made with a clean scalpel.

d. Place clippings and scrapings in a clean vacutainer, seal, and label.

e. Refrigerate, but do not freeze. This will prevent putrefaction of blood, etc.

f. Transport to forensic laboratories as soon as possible.

g. Comparison standards — may be retrieved from a suspect, making sure that proper (legal) procedures are followed.

12. GLASS

a. Glass fragments seized may provide a source for physical matching. In addition to physical matching, various types of glass may be identified as coming from a similar source.

b. Package glass fragments in leakproof plastic vials.

c. Seal and mark container, noting from where sample was taken.

d. Large chips or samples should be double wrapped.

e. If a physical match is sought, all edges should be carefully protected from further damage.

f. Comparison standard: any pane of glass from a window, vehicle, drinking glass, bottle, or headlight of a car (hit and run).

g. The suspected source (bottle, headlight, etc.) should be seized in its existing form and carefully packaged for transportation.

13. HAIR

a. Collect as you would a fiber (see Fiber).

b. Either package the item or remove hairs with clean tweezers and place in a sterile container, seal, and label.

c. If collected from victim (either live or at autopsy), do not cut.

d. Scalp hair: Comb first with clean comb, and then pull hairs from various locations on the scalp. Collect a minimum of 25 total hairs from the front, back, and both sides of the scalp.

e. Pubic hairs: Comb the area first, then pull a minimum of 20, preferably 50, from the area if possible. Ensure pulled hairs are packaged separately from combed hairs.

f. Other body hairs: Comb or pull a minimum of 20 hairs from the site of the chest, limb, beard, mustache, etc.

g. Comparison standard: All hairs must be compared to both the victim and the suspect. Always package all hair separately in sterile or clean containers, seal, and label.

14. HANDWRITING/ HAND PRINTING

a. Both written and printed documents may be favorably compared with a known source (person). Handwriting comparison may be advisable when a note is left at the scene of a suicide to ensure that it was written by the victim.

b. Air dry all documents prior to packaging.

c. Place documents between layers of plastic or paper.

d. If documents are to be examined for latent fingerprints, this should be done first, with the latent print examiner being advised that a handwriting examination will be conducted later. This will let the latent print examiner know not to use chemical processing techniques that contain solvents that will dilute or destroy the inks used on the document.

e. Comparison standard: Should be as extensive as possible. Ideally, comparison standards are written using the same type of writing instrument under similar conditions. Obviously this may not be possible when the exhibit is a suicide letter and all comparison standards have been written in the past. Collect as great a quantity of the victim's handwriting samples as is possible, package, label, and forward with the known sample for comparison. The comparison standards should (ideally) be witnessed documents to ensure that they were indeed written by the victim.

15. LIVER

a. During the autopsy, obtain at least 200 grams (4 ounces) of liver tissue and place the sample in a clean, leakproof container. Seal and label.

b. Refrigerate, do not freeze.

16. LUBRICANT

a. When lubricant is suspected on an undergarment, air dry the garment and seal in a clean container or plastic bag.

b. Comparison standard: Possible standards (exhibits) may be located in the victim's OR the suspect's undergarments/clothing (e.g., Vaseline, etc.), vehicle or residence.

17. LUNGS

a. Samples taken during the autopsy may provide clues as to cause of death. These clues include solvents (glue, PAM, cleaning fluids), carbon/ smoke particles (when death was due to smoke inhalation), diatoms or bottom debris (where drowning is suspected), or any other material normally foreign to the lungs.

b. Place lungs in a leak-proof, airtight container, seal and label.

c. Refrigerate, do not freeze.

d. Transport as soon as possible.

18. MATCHES (PAPER)

a. Carefully wrap each match in clean tissue paper or soft material, and place in a protective container to avoid any damage to the torn end.

b. Comparison standard: A match found at the scene may be matched to a specific book of matches found elsewhere (e.g., in a vehicle, on a suspect, etc.).

19. PAINT

a. Paint samples taken at a crime scene are particularly valuable in crimes that involved hit and run or death due to blunt trauma injury.

b. Samples may be taken from skin, clothing, vehicles, or even objects such as rocks or any object that could be used as a club or weapon.

c. Do NOT use adhesive tape for lifting/removing paint from an object. Use a clean scalpel or knife whenever possible. A disposable razor blade is ideal for this type of evidence collection and should be used for only one sample.

d. Place particles of paint in a clean (preferably glass) container.

e. Protect the edges of paint chips when physical matching is to be considered. In cases where physical matching is considered, the chips must be carefully sandwiched between layers of soft, clean material to protect further damage.

f. Comparison standard: Paint chips may be physically matched to their source in the case of hit and run, motor vehicle accidents, or chemically compared to any other source when only very small particles are recovered.

g. Submit paint samples in original paint tins, when the comparison standard is suspected to be a liquid.

h. All samples should be packaged separately to avoid cross-contamination.

20. PHYSICAL MATCH

a. Any object may (potentially) be physically matched to its source. This includes items such as pieces of rope, string, wood, broken bone, paper (torn), tape, etc.

b. Package all "physical match" items carefully to avoid any damage to torn or broken ends that are to be matched to a known source (comparison standard). Secure packaging to avoid further damage or alteration of the torn/broken end is of paramount importance.

21. PILLS

a. Pills should be kept in their original container.

b. When found loose, they should be placed in a clean vial and secured.

c. After packaging, sealing and labeling, they may be transported for analysis or comparison to known standards. The standard *Compendium of Pharmaceuticals and Sundries* identifies (visually) most prescription and nonprescription pills on the market in the United States of America and Canada.

d. Pills may also be compared to stomach contents, liver, etc.

22. SALIVA

a. Saliva may be compared to a known blood group or DNA match in some instances.

b. Saliva stains should be air dried completely. Package the stained item in a sterile container, seal and label. If completely dried, the exhibit may be left at room temperature until transported and examined.

c. Comparison standard: Comparison is made with a blood sample taken from the suspected source (individual).

23. SEMEN

a. If the goal is merely to identify the substance as semen, a swab of material (clothing) may be air dried, packaged in a sterile container, sealed and labeled prior to shipping. If blood is not suspected as forming part of the stain, the exhibit may be stored (after air drying) at room temperature. If the investigator suspects that seminal fluid may be present, he can use an ultraviolet light as a scanning tool to locate and detect substances that may be semen. He should then collect the glowing area or object for further analysis at the laboratory.

b. If the goal of the collection is blood typing or DNA matching, the substance/stain may still be air dried at room temperature, but must be refrigerated after packaging.

c. Comparison standard: Semen may be compared to any individual; boyfriend, husband, or any other male.

24. STOMACH CONTENTS

a. During the autopsy, collect as great a quantity of stomach contents as possible.

b. Place in clean, airtight container, seal and label.

c. Refrigerate, but do not freeze.

d. Comparison standard: Stomach contents may yield information as to the contents (and time) of last witnessed/documented meal as well as any medication/drug/poison.

25. TAPE

a. Tape may be physically matched to its origin. It may also retain fingerprints on the shiny surface or on the sticky underside.

b. Tape should be carefully packaged, ensuring the ends are protected. If the tape is still stuck to an object, the entire object (or the portion of the object adhering to the tape) should be seized as an exhibit. It should be removed only by a qualified technician capable of proving a match or examining for fingerprint impressions.

c. The sticky side of tape or the sticky edge of a roll of tape may contain hair or fibers, especially when the roll has been carried in a pocket.

26. TIRES (PUNCTURED OR SLASHED)

a. It is not necessary to submit the entire tire for examination.

b. Using a sharp instrument, cut the area around the suspected puncture or slash, leaving several inches of material, package carefully and transport.

c. Tires that may be the subject of a comparison to a suspected tire impression should be collected and retained for a comparative analysis.

27. TOOL MARKS

a. Tool marks (pry marks, etc.) may be physically matched to a specific tool using a comparison microscope.

b. The object bearing the tool mark impression should be submitted. When this is not possible, it should be carefully removed from the parent source and packaged softly and securely to avoid any alteration or disfiguration. Casting material such as Mikrosil may be used if the object bearing the tool mark cannot be removed.

c. The impression must be kept separate from the suspected tool.

d. Comparison standard: any tool, prying instrument or any object that could have made the impression.

28. TOOLS

a. Tools may be matched to specific impressions, pry marks or (on rare occasions) indentations.

b. Package carefully to protect the end of the tool. Use plastic bags or Styrofoam cups taped over the end of the tool when it is suspected that there may be material impacted on the end of the tool.

c. Transport in a secure, tightly wrapped container to prevent movement.

d. Keep all tools separate from suspected tool marks.

e. Refer to the headings of Glass, Paint, and Tool Marks.

29. UNKNOWN LIQUID OR POWDER

a. Any unknown liquid or powder should be submitted in its original container, if possible.

b. Seal all liquid samples to prevent leakage, evaporation or contamination.

c. Label.

d. If liquid is suspected of being volatile, ship according to federal laws governing the transportation and shipment of hazardous goods.

e. Unknown liquids or powders may be submitted to determine their identity (e.g., alcohol, drugs, poisons, etc.).

30. URINE

a. All urine exhibits should be refrigerated (not frozen) until transported to a forensic laboratory for examination.

b. Urine may be obtained during the autopsy and placed in a clean leakproof container.

c. Urine may be analyzed for the presence of alcohol, drugs, or poison.

d. Ideally, urine should be stored in sterile, clean vacutainers or urine specimen jars..

31. VITREOUS HUMOR

a. Vitreous humor (the clear fluid contained within the eyeball) may be collected and analyzed for alcohol content. This is a particularly useful technique when a quantity of blood cannot be obtained. Vitreous humor must be collected during the autopsy, not from a living person

b. Collect one vacutainer vial XF947 or 10 ml. gray stoppered vacutainer. Mix gently. Seal stopper. Label.

c. Refrigerate, do not freeze. Transport to forensic (alcohol) laboratories as soon as possible.

A GLOSSARY OF TERMS

14

In the making of this manual, an attempt has been made to refrain from technical jargon. The purpose of this manual is to reach the public safety diver, and it is understood that a great many individuals involved in this profession are not full-time sworn or trained law enforcement people. For this reason, this listing of commonly used terms and their definitions is included in this manual.

The terms defined in the following pages commonly appear in police reports, coroner's reports, and pathologist's reports. The public safety diver should not only be able to define each term, but also fluently use them in the body of his investigative report.

Although this is not a comprehensive list, it is a list of the most commonly used terms in this field. These terms should also be included as a part of the language used by the public safety investigative dive team.

Acid phosphatase — An enzyme group found in many parts of the body. It is only found in significant concentration in the prostate gland. The presence of acid phosphatase on clothing or in vaginal washings indicates the presence of seminal fluid. Acid phosphatase may be removed from clothing that has been submerged in water. The presence of acid phosphatase, even in the absence of sperm cells, is corroborative evidence that sexual intercourse has taken place.

Adipocere (Adipocire) — A white-gray waxlike substance formed when free fatty acids within the body undergo hydrolysis in a moist alkaline environment. The presence of adipocere inhibits further decomposition.

Adsorb — To gather or condense on a surface, including interior, porous surfaces such as that of activated charcoal.

Adult teeth — The teeth that replace the deciduous teeth. The adult teeth are often referred to as the permanent teeth. The first adult tooth usually erupts around the age of six years.

Agonal — Relating to the last moments of life or periods of great pain or misery (e.g., agonal struggle of a drowning victim or agonal cries of a tortured person).

Air embolism — A blockage within the circulatory system due to the introduction of air into the arteries or veins. In scuba diving, the air may be forced directly into the blood vessel network surrounding the lungs by lung overexpansion and subsequent bursting of the alveoli. The event is also referred to as burst-lung syndrome or arterial gas embolism (AGE).

Algor mortis (Algormortis) — Postmortem body cooling.

Alveoli — Tiny air sacs that are found at the terminus of each respiratory bronchiole. The alveoli make up the structure of the lungs, with the lung structure often being compared to a bunch of grapes, the alveoli being the grapes and the respiratory bronchioles being the stem leading to each grape. The number of alveoli in each lung numbers in the billions. Singular: alveolus.

Amalgam — An alloy of silver and tin, mixed with mercury. This is commonly used in the restoration of posterior teeth (dental fillings).

Ambient — Surrounding, in the immediate vicinity.

Anoxia — A complete lack of oxygen, without oxygen.

Ante Mortem (Antemortem) — Before death (*ante = before; mortem = death*)

Anthropophagy — The eating of human flesh. Anthropophagy includes scavenging or carrion-eating, predation, and cannibalism.

Aorta — The single main artery carrying blood away from the heart. The aorta exits the top of the heart, then branches out as it curves downward as well as upward to all parts of the body. The single greatest artery of the body.

Apnea (Apnoea) — Without breathing (e.g., voluntary breath holding is voluntary apnea).

Arrhythmia — Without rhythm. A condition of the heart where the organ beats sporadically or in a nonconsistent manner is said to be a condition of arrhythmia. Arrhythmia usually precedes cardiac arrest.

Arteriosclerosis — Hardening of the arteries. A disease that is often associated with heart and kidney disease. Arteries lose their elasticity and become narrowed due to deposits laid down on the inside walls. Common among the aged and obese. Arteriosclerosis is commonly thought to be accelerated by smoking and improper diet.

Articulated — Refers to skeletal remains that are still connected by soft tissue. Also refers to skeletal remains that have lost all soft tissue yet still remain in their proper anatomical position.

Atlas — the first cervical vertebra, which is located at the top of the spinal column adjacent to the cranium.

Atypical — Not typical.

Autolysis — The breaking down of cells by endogenous enzymes. Autolysis is usually the first structural postmortem change in a cadaver. The term autolysis is loosely translated as "self-digestion."

Azoospermia — A complete absence of sperm cells.

Barotrauma — A physical injury due to the direct effects of pressure change. Most often barotraumatic injuries include the middle ear, sinus cavities or lungs. Barotrauma has been documented in sinuses or middle-ear cavities of drowning victims in as little as 6 feet of water.

Bends — A disease common among men who work in an environment having greater than atmospheric pressure. Originally so named because of the long-term crippled or bent spines left in those who had been exposed to this disease repeatedly. Also called caisson's disease or decompression sickness.

Bilateral — Both sides.

Biological identity — Refers to the basic biological characteristics of all persons in terms of age, sex, race, and stature.

Bronchioles — Small air passages within the lungs. The smallest of these air passages lead directly to the alveoli and are less than the diameter of a human hair.

Cadaver — A dead body; a corpse.

Cadaveric spasm (Cadavaric spasm) — A muscular spasm occurring at the moment of death and remaining afterward. Unlike rigor mortis, it is confined to one group of voluntary muscles and is sometimes observed in drowning victims clutching bottom debris or personal belongings. Also referred to as instantaneous rigidity and cataleptic rigidity.

Caisson's disease — See Bends. So named because of the pressurized chambers or "caissons" construction personnel worked in when tunneling beneath rivers, etc. Also called decompression sickness.

Calcaneus — The largest of the ankle (tarsal) bones. The calcaneus forms the heel of the foot.

Calliphora vomitoria — A common fly often referred to as the blow fly or blue bottle. They are noticeably larger than the common housefly.

Canine — Also known as the eye tooth or cuspid. The third tooth from the front, characterized by its sharp pointed appearance and long root.

Capitate — The largest of the eight wrist (carpal) bones.

Cardiac — Referring to the heart.

Carotid sinus — A small organ located in the wall of both carotid arteries. It is responsible for regulating blood pressure.

Carpal/carsus — The wrist and/or the wrist bones as a whole. There are eight carpal bones forming the wrist and are structured in two rows of four.

Catharsis — A sudden emotional outpouring. A sudden release of previously repressed or suppressed feelings.

Caucasoid — A major racial stock of humankind, originating in European ancestry. Usually but not always synonymous with "white." This assumption may be incorrect.

Causa causans — (Latin) The immediate cause.

Center of ossification — The areas within a body or within a soft tissue that precede the formation of bone. The location where bone initially forms. Many bones have several centers of ossification. For example the center of ossification for a long bone is primarily located at both ends.

Central nervous system — Referring to the brain.

Cervical — The region of the neck, specifically the seven vertebrae of the neck.

Chloride test (Gettler test) — A postmortem blood test used to determine whether or not death was due to drowning. This test has recently been found inconclusive and at times quite misleading. It is still used but remains highly controversial.

Chronologic age — The actual or calendrical age. Time from birth to present (in a living person) or death (in a deceased person).

Clavicle — Commonly referred to as the collar bone. The clavicle is a common site of fracture and often bears evidence of pre-, peri-, and postmortem trauma.

Coccyx — The tail bone or lowest bone of the spine. The coccyx consists of three to five bones.

Colleoptera — An order of insects characterized by biting mouth parts, membranous hind wings beneath a thick, tough outer wing, and complete metamorphosis. Common name: beetle.

Congestion lividity — Congestion of surface blood vessels with blood. This congestion is due to gravity causing the blood to flow to the lower areas of the cadaver. It is exhibited as a dark coloration of the skin, often mottled.

Convolution — Ridged. Often referring to the surface of the brain.

Coracoid process — A small fingerlike process or projection on the scapula.

Cornu — A small projection on the hyoid forming the arms of the "U" from which ligaments attach the hyoid to the base of the skull. Plural: cornua.

Corpus delicti — The main part or the body of a crime. Corpus delicti does not refer to the corpse involved in a homicide, although it may be used in reference to an object (possibly a body) central to the crime, which may be used as evidence.

Costal — Refers to the ribs. For example, intercostal refers to the space between the ribs.

Cranium — The portion of the skull excluding the lower jaw and hyoid.

Cratering — This term describes the effects of a projectile (bullet) on bone where the exit hole is larger than the entry hole due to an expanding bevel (cone of percussion). An example of this effect is often seen in glass where a bullet has penetrated a window without shattering the pane.

Cuboid — One of the seven ankle (tarsal) bones. The cuboid is located on the outside or little-toe side of the ankle.

Cuneiform — A wedge-shaped bone in the ankle located between the inside big toe of the foot and the cuboid.

Cutaneous — Pertaining to the skin.

Cutis anserina — Commonly called goose-flesh or duck-bumps, etc. It is a bumpy appearance of the skin caused by contraction of the erector muscles at the base of the hairs. Most often observed on the arms, legs, chest or back. In a living person it is caused by fear, cold, etc. In a cadaver, it is as a result of rigor mortis in these small muscles. It has no bearing on whether or not death occurred in the water.

Cyanosis—A bluish discoloration of the skin caused by insufficient oxygen supply to the body. The coloration is supplied by carbon-dioxide-laden hemoglobin (in red blood cells — only oxygenated hemoglobin is red, unless carbon monoxide is present).

Decedent—Deceased person; cadaver.

Deciduous—Those teeth that are first to erupt and later replaced by the permanent teeth.

Defense wounds — Injuries usually observed on the forearms or palms of the victim. These wounds are commonly inflicted (by a knife, etc.) during a struggle.

Dental restoration — Refers to the process and materials involved in reconstructing teeth or repairing a cavity. Common materials include (but are not restricted to) gold, porcelain, silica, or plastic resins. The latter three present problems when dental comparison X-ray is utilized as a means of identification since they show up only as faint shadows on X-ray film.

Dentition — Refers simply to a set of natural teeth.

Diatom — A single-celled organism whose skeleton is composed mainly of silicon dioxide. Presence of diatoms in the bone marrow of a cadaver may lead to speculation that death was due to drowning. Recent studies have shown this to be nonconclusive. A more positive determination would be to compare diatoms found within the victim's lungs or bone marrow with diatoms taken from a water sample retrieved from the location where the body was found. There are millions of varieties of diatoms found throughout the world.

Dilate — Widen or expand. Dilated pupils are pupils that are large or expanded.

Diptera — An order of insects characterized by a single pair of functional wings, sucking mouth parts, and complete metamorphosis. Common name: fly.

Distal — Furthest from the trunk of the body. The hand is distal to the elbow.

Drowning — Death due to the immersion of the nose and/or mouth in water or any liquid.

Dry drowning — A condition of drowning where very little water has entered the lungs.

Dyspnea — Shortness of breath, having difficulty in breathing.

Edema — The presence of excess fluid in the tissues. In drowning victims, edema of the lung tissue is common.

Emaciated — Lean, wasted away.

Embolism — A condition existing when a blood vessel is clogged (with an embolus). The clotting material may be composed of a blood clot, fat or any other substance. See also: air embolism.

Embolus — The clot in the condition referred to as embolism. Plural: emboli.

Empyema — Presence of pus within a body cavity, usually within the pleural cavity.

Epiglottis — A thin cartilage located at the base of the tongue that covers the airway during swallowing.

Epiphysis — A secondary center of ossification located at the end of growing bones. It is separated from the primary center by a tough, fibrous band of soft tissue. After prolonged decomposition, the fibrous union will decay, yielding a loose epiphysis that can be easily overlooked and lost during the recovery of immature skeletal remains. Plural: epiphyses.

Esophagus — The muscular tube that transports what is swallowed (bolus) to the stomach.

Eversion — The degree of outward flare at the angle of the lower jaw. The eversion is more pronounced in males and heavy chewers.

Exine — The outer covering of pollen grains.

Exitus — Death.

Exposure cracking — Most often observed on the cranium. Exposure cracking is caused by repeated bouts of exposure of the skull to moisture and high temperatures. When this occurs, stress cracks are often observed on the outer surface of the bone.

Exsanguination — Death due to the loss of blood from the circulatory system. Bleeding may be internal or external.

Extended burial — Refers to the position of a body in which the lower limbs are straight in line with the trunk. The position of the arms may vary.

Feces — Excrement or stool. Fecal matter may be collected for analysis of foods or drugs taken before death. Animal feces are called scats.

Femoral — Referring to the femur.

Femur — The largest limb bone forming the upper leg; thigh bone.

Fetal — A developing human between four months intrauterine (since conception) and full term.

Fibrillation — Fibrillation of the heart means an insufficient beating or fluttering. A condition where the heart muscle is contracting out of synchronism, resulting in no (or very little) blood being pumped through the body.

Fibula — The long flat outer bone of the lower leg.

Flaccid — Flabby, soft to the touch.

Flexed burial — Refers to the position of a body, usually lying on its side with the limbs drawn up into a contracted position. Such positioning would normally indicate a deliberate positioning of the limbs by a second party except in the case of a newborn infant.

Foramen — A passage or hole in a bone or organ. These holes or tunnels occurring naturally in bone allow the passage of nerves and blood vessels. Plural: foramina.

Foramen magnum — The large hole in the base of the skull through which the spinal cord connects to the medulla oblongata (brain stem).

Foramen ovale — The hole in the middle of the heart that is usually closed or sealed at birth. When this hole is not sufficiently closed, insufficient oxygenation of the blood results in what is commonly referred to as a "blue baby." See: cyanosis.

Forensic — Dealing with courts of law. Scientific tests or techniques used in connection with the detection of crime.

Forensic anthropology — The study of human bones and teeth used to determine the identity or identifying details of human remains as well as interpretation of circumstances affecting the death of the individual and postmortem interval.

Forensic medicine — Medical knowledge applied to criminal investigation or any legal concern.

Fusion — The replacement of soft tissue between two bony elements by bone. This results in a solid union (fusion). The maturing of a human skeleton is characterized by progressive fusion that reduces the number of individual bone elements.

Gamete — A general term often used to refer to the male or female reproductive cell. Also often used to refer to pollen grains.

Gastrointestinal — Pertaining to the digestive system (esophagus, stomach, intestines).

Glabella — A landmark on the skull directly beneath the normally appearing bald spot between the eyebrows.

Glia — Supporting tissue of the brain.

Glioblastoma multiforme — A type of brain tumor.

Glioma — A type of brain tumor.

Glottis — The opening at the top of the windpipe.

Growth arrest line — Also known as Harris' lines. Dense bands of bone near the extremity of a skeletal element. These lines mark the former end of the growing bone when its growth is temporarily arrested due to illness or a period of malnutrition. These lines may assist in individual identification of an individual when their past health is well documented.

Grub — Larval form of coleopteran insects.

Gyrus — A convoluted ridge of the brain. Plural: gyri.

Hard tissue — A term used to describe bones and teeth.

Hemopericardium — Blood in the sac surrounding the heart.

Hemopneumothorax — Presence of blood and air within the chest cavity.

Hemoptysis — Coughing or spitting up of blood.

Hemorrhage — The escape of blood from a blood vessel.

Holometabolous — A term used to describe insects that exhibit dramatic growth transformation from egg to larva to pupa to adult. During these transformations, they undego metamorphosis involving shape and size (i.e., beetles and flies).

Humerus — The long bone of the upper arm.

Hydrostatic test — A test used to determine live birth by observing buoyancy of the lungs (of the baby) in water. Also referred to as the flotation test. Also, a test of high pressure cylinders to determine the degree of metal fatigue.

Hyoid — A U-shaped bone on the floor of the mouth supporting the tongue. A fractured hyoid may be indicative of strangulation.

Hyperbaric — Increased pressure. When the human or animal body is the object of study, the term "hyperbaric medicine" is used.

Hypersensitivity — An increased sensitivity, beyond the normal range.

Hypostasis — A settling of the blood due to depressed circulation.

Hypoxia — Lack of sufficient oxygen.

Incisor — Teeth that are located at the front of both the upper and lower jaws. Characterized by a chisel-like crown and single short root.

Incus — One of the three very small bones located in the middle ear. Common name: anvil.

Infarct — An area of dead tissue caused by blockage of blood to that area. Coronary infarct is a common type of heart attack.

In mortuo — In the dead body.

Innominate — The hip bone or pelvis. It is formed through the fusion of the ilium, ischium, and pubis at puberty.

In situ — In place.

Instar — A developmental stage of an insect larva.

Intrauterine — Synonuymous with in utero.

In utero — In the womb or uterus. An age of a fetus is often expressed in weeks or months since conception (i.e., 20 weeks in utero).

In vivo — In the living body.

Lactic acid — A substance produced by respiration. Lactic acid is produced in muscle tissue when there is insufficient blood flow to carry away abnormally high levels of carbon dioxide. In a corpse, it is instrumental in producing rigor mortis.

Lanugo hairs — Hair appearing on the human fetus during the seventh and eighth month of gestation. These hairs are almost gone at birth. Presence in the lungs demonstrates amniotic fluid embolism.

Larva — Pre-adult developmental stage derived from an insect egg prior to metamorphosis. The larva of a common fly is commonly referred to as a maggot. The larva of a beetle is likewise called a grub.

Laryngeal spasm (laryngospasm) — An uncontrollable spasm (tightening) of the muscles surrounding the larynx. Usually caused by the introduction of a foreign object (water, etc.).

Larynx — The portion of the airway containing the voice box or vocal cords.

Lingual — Indicates the surface of oral structures (most commonly teeth) that face the tongue.

Livor mortis — Pooling of the blood after death. This pooling is visible as a darkened skin coloration and is due to gravity.

Lumbar — The five vertebrae in the lower back region that do not bear ribs.

Malleus — One of the three very small bones of the middle ear. Common name: hammer.

Mandible — The lower jaw. At birth, the mandible is separated at the chin into left and right halves.

Mandibular torus — A raised series of dense bony bumps or a fairly continuous ridge of bone situated on the lingual (tongue) side of the lower jaw, below the teeth. This is most often seen in North American native peoples.

Marbling — Postmortem vascular (blood vessel) patterns appearing on the skin.

Mastoid process — A bony projection that can be felt just behind the ear lobe. This is usually larger in males.

Maxilla — The upper jaw.

Medial — Designates the surface of a structure which faces toward the midline of the body. For example, the medial side of the thigh would be the inside.

Meninges — A collective name given to the three membranes surrounding the brain. They are the pia, arachnoid, and dura.

Metacarpals — The five bones forming the palm of the hand between the fingers (phalanges) and the wrist.

Metacarpus — General area included by the metacarpals.

Metamorphose — The act of undergoing an extreme change, i.e. the transition from a caterpillar to a butterfly. Also referred to as metamorphosis.

Metatarsals — The five bones of the foot between the toes and the ankle.

Milk teeth — Also known as the deciduous teeth. There are five milk teeth in each quadrant of the mouth, 20 in total.

Modus operandi — The method. Usually referring to the method of committing a crime. Commonly referred to as the "M.O."

Molar — Chewing teeth located at the back of the mouth. Molars have two or three roots and a crown with a large chewing surface. Normally there are two deciduous and three permanent molars in each quadrant of the jaw.

Mongoloid — A major racial origin believed to originate in Eastern Asia. Mongoloids share common features such as yellow to brown skin and straight black hair.

Myocardium — Pertaining to the heart muscle (e.g., myocardial infarct means an interruption of blood supply to a portion of the heart muscle, resulting in the death of the cells concerned). Myocardial infarct is a common form of heart attack.

Nasal sill — The bony structure projecting from the lower margin of the nasal opening. This is the location where the cartilage forming the nostrils attaches to the midline.

Necrosis — Death of a tissue without necessarily causing death of the organism.

Negroid — A major racial category of humankind largely of African ancestry. The negroid classification shares a complex of physical features such as course hair and dark skin.

Nuchal lines — Ridges of raised bone on the lower part of the back of the cranium where the neck muscles are attached. Often more pronounced in males.

Occipital — A bone at the back of the cranium to which the neck muscles are attached.

Occlusal — The chewing surface of the teeth.

Ocular (occular) — Pertaining to the eye.

Orbit — The bony socket of the eye.

Order — A large scale grouping of biological organisms all sharing a number of physical and behavioral features. For example, the order Primates includes monkeys, apes, and man.

Ossification — The process of the formation of soft tissue into bone.

Osteology — The study of bones and skeletons, including their evolution, growth, and function.

Palate — The roof of the mouth

Palmar—Pertaining to the palm of the hand.

Palynology — The study and analysis of pollen grains, including their size, shape, and origin. Identification of pollen grains found on or in a body may assist with determining the location and/or season in which the individual died.

Parietal — The two bones that meet at the midline of the cranium.

Patella — The knee cap.

Pathology — Medical knowledge relating to the alteration of tissues due to disease, aging, violence, or death.

Perforating wound — A wound that enters and passes all the way through to the other side of the body (or any anatomical structure). The importance of a perforating wound is that the holes of entry and exit must not be confused with two separate and distinct woundings.

Pericardium — The sac surrounding the heart.

Perimortem — Around the time of death. Perimortem wounds usually are those that occur within five to 10 minutes before or after death. These are commonly mistaken for postmortem wounds. A perimortem wound may be critical evidence in ascertaining the cause of death.

Perinatal — Around the time of birth.

Permanent — When used to describe teeth, refers to the adult teeth.

Petechia — Tiny, spotlike hemorrhages.

Petrous — A dense bone located at the base of the cranium. The petrous contains the three ear bones (ossicles): the incus, malleus, and stapes. The petrous bones is one of the first in the human body to ossify and usually does so in the fetus at approximately five months in utero.

Phalanges — The bones of the fingers and toes.

Phlebothrombosis — The presence of a clot (thrombus) within a vein.

Pisiform — A pea-sized wrist bone.

Plankton bloom — A condition experienced when specific types of plankton multiply in a "greater than normal" fashion. Plankton blooms occur both in lakes and oceans. They are usually short-lived and result in a temporary clouding of the water.

Plantar — Pertaining to the soles of the feet.

Pleural — Pertaining to the lungs or lung cavity.

Pneumothorax — The presence of air within the pleural (lung) cavity. This may occur as a result of a tear in one or both lungs or an external puncture wound in the chest.

Postmortem interval — The time lapse between the moment of death and the examination of the cadaver.

Postmortem (post mortem) — Present or occurring after death. When used as a noun, it refers to the autopsy, properly called a postmortem examination.

Postmortem suggillations — Blood pooling after death. Postmortem lividity.

Premolar — Adult teeth that replace the deciduous molars. The premolars are located between the canines and the molars. There are two in each quadrant and are characterized by having (usually) one root and two cusps (peaks) on the occlusal (biting) surface.

Premortem — Prior to death. Also called antemortem.

Primary ossification center — The site where a future bone starts to ossify, prior to the appearance of a secondary center of ossification for the epiphyses. In a major long bone (i.e., femur) the primary ossification center grows to form the shaft while the epiphyses form at either end.

Process — Any growth of bone that projects from the main portion of the structure; i.e., the spinal processes are the lumps that can be felt down the posterior surface of the backbone.

Pronated — The position of the forearm when the hand is rotated palm-down.

Prosthesis — Artificial appliances used to replace missing or removed body parts. Includes artificial limbs, eyes, teeth, etc.

Protoplasm — The portion (contents) of a cell outside the nucleus and inside the cell membrane.

Proximal — Close to the trunk; i.e., the shoulder is the proximal end of the arm.

Pulmonary — Relating to the lungs.

Pupa — Growth stage of an insect between larva and adult.

Pupation — The process of becoming a pupa.

Purpura (pupuria) — Irregular hemorrhages occurring in some mucus membranes and the skin.

Putrefactive gases — Gas produced in the cadaver after death.

Radius — One of the two bones of the forearm, located on the thumb side.

Respiratory bronchioles — The smallest of airways contained within the lungs. These lead directly to the alveoli.

Rigor mortis (rigormortis) — Stiffening of the muscles after death. Rigor mortis is a transient phenomenon.

Sacrum (sacra) — The five fused vertebrae at the base of the spine are collectively referred to as the sacrum. The bone is situated within the pelvic girdle which supports the spinal column and the weight of the chest and head. The sacrum is often useful as an indicator of sex.

Sanguineous — Possessing the color of blood.

Saponification — The process by which body fats are converted to a greasy soaplike substance. See: adipocere.

Saprophagous — An organism (insect, crustacean, etc.) that feeds on rotting flesh and/or bodily fluids.

Scapula — The shoulder blade.

Scat — Animal droppings or feces.

Sebaceous — Containing fat or oily matter. A type of latent fingerprint impression that may remain even after submergence in water.

Secondary ossification center — Or epiphysis, a bony center that grows at the ends of a major bone. Secondary ossification centers may ultimately fuse, resulting in the cessation of growth.

Shapiro plateau — See: Temperature plateau.

Sinus — An air-filled cavity within a bone. Any designated cavity within a structure, however, may be referred to as a sinus.

Slim Jim — A tool used to open locked car doors. A Slim Jim is made from flat spring-steel and is usually cut or notched on one end. It is forced down the outside of a car (side) window and used to manipulate the locking linkage to open the vehicle's door without the use of a key. In some countries, the possession of a Slim Jim without legal justification constitutes a criminal offense.

Sounding line — A thin line marked off in increments (usually 10-foot/3-m intervals) and weighted at one end. Its primary use is for determining water depths. Braided nylon cord, one-eighth inch/3 mm diameter and a two-pound (~ 1 kg) weight are common choices.

Stapes — One of the three small bones of the ear. Common name: stirrup.

Sternebrae — The sternum (breast bone) is composed of several (usually five) separate bones, or sternebrae, which become fully fused around puberty.

Sternum — The breast bone, composed of the manubrium and sternal body.

Stokes-Adams Syndrome (Stokes-Adams' syndrome) — Sudden unconsciousness. Sometimes due to heart attack.

Subcutaneous — Pertaining to the area immediately beneath the skin.

Supramastoid crest — The bony ridge located just above each ear hole. Often more pronounced in males.

Supraorbital ridge — The raised ridge of bone above each orbit (eye), usually felt at or slightly above the eyebrow. More pronounced in males.

Syndrome — A set of signs and symptoms occurring simultaneously.

Talus — A major ankle bone between the heel and the lower leg bones.

Tardieu's spots — Spotlike hemorrhages, once thought to be associated with asphyxia.

Tarsals — The seven ankle bones of the foot.

Temperature plateau — Also referred to as the Shapiro plateau, named after the individual who first documented the phenomenon. Pertaining to the period immediately following death, in which the core temperature of the cadaver does not fall. Due to many factors, this may last from one to five hours. See also: virtual cooling time.

Tibia — The major bone of the lower leg; i.e., the shin bone.

Trapezium — A small wrist bone.

Trapezoid — A small wrist bone.

Travel abrasion — Refers to the postmortem injuries to the cadaver caused by movement along the bottom of a body of water. It may be caused by current or wave action.

Triquetrum — A small wrist bone.

Ulna — One of the two bones of the forearm located on the side opposite the thumb.

Vade mecum — A handbook or reference book to be carried as a ready reference manual.

Valsalva maneuver — Pertaining to a voluntary increase of pressure in the lungs. This is effected by attempting to breathe out against a closed nose and mouth, as when "popping" one's ears, or against a closed glottis. Repeated performance of the Valsalva maneuver may result in irregular heartbeat or death.

Vascular — Referring to the blood vessels.

Vehicle identification number (VIN) — A serial number placed on a vehicle by the manufacturer. There are usually more than one, with a "hidden" or "secret" number also being registered with the manufacturer. The location of the secret VIN can be obtained from the manufacturer and varies with the different makes, models, and vintage.

Ventral — A surface facing toward the front of the human body; i.e., the face is on the ventral side of the head.

Ventricular tachycardia — A speeding or racing (increased rate of beating) of the heart, in particular the ventricles.

Vena cava — The largest vein in the body, responsible for returning blood to the heart.

Vertebra — The segments of the spine.

Vertigo — Dizziness. Alternobaric vertigo is a (often) severe dizziness experienced by divers, usually on ascent. It is caused by a pressure or temperature difference in the semicircular canal (organ of balance located in the ear).

Virtual cooling time — The time taken for a cadaver to cool (core temperature) through the first 85 percent of the difference between its original (antemortem) temperature and the ambient temperature. The virtual cooling time is often utilized to determine the postmortem interval.

Viscera (visceral) — Pertaining to the space within the main body cavity. The space containing the major body organs.

Visual inspection — The internal examination of a diving cylinder with a light or other device, usually conducted at intervals of one year or less, and noted by affixing a sticker with the inspection month and year to the exterior of the cylinder.

Vital reaction — A reaction in a living tissue such as inflammation or hemorrhage. It is often used to determine whether a wound was antemortem or postmortem.

Vitrification — When bones are heated to temperatures in excess of 1,500°F (800°C), they may acquire a hard porcelainlike quality. In addition to this, they may ring when tapped against a hard surface.

Vomer — A small bone located behind the nose at the midline.

Wauschaut (washerwoman's hands) — The wrinkling of the skin of the hands or feet caused by prolonged exposure to moisture. Wauschaut may occur before, during, or after death.

Wet drowning — A drowning that occurs when large quantities of water (or any other fluid) are aspirated directly into the lungs.

Zygoma — The cheek bone. More pronounced in Mongoloid peoples.

CPSIA information can be obtained
at www.ICGtesting.com
Printed in the USA
LVHW050057300622
722367LV00008B/215

9 781930 536722